A GUIDE TO THE STUDY OF environmental pollution

contours: studies of the environment

Series Editor
William A. Andrews
Associate Professor of Science Education
The College of Education
University of Toronto

A Guide to the Study of ENVIRONMENTAL POLLUTION
A Guide to the Study of FRESHWATER ECOLOGY
A Guide to the Study of SOIL ECOLOGY
A Guide to the Study of TERRESTRIAL ECOLOGY

A GUIDE TO THE STUDY OF

environmental
pollution

Contributing Authors:
William A. Andrews
Donna K. Moore
Alex C. LeRoy

Editor:
William A. Andrews

Prentice-Hall of Canada, Ltd., Scarborough, Ontario

A Guide to the Study of
ENVIRONMENTAL POLLUTION
© 1972 by W. A. Andrews
Published by Prentice-Hall of Canada, Ltd.,
Scarborough, Ontario.
Printed in Canada.
ISBN 0-13-370833-0   1 2 3 4 5 76 75 74 73 72

Prentice-Hall, Inc., *Englewood Cliffs, New Jersey*
Prentice-Hall International, Inc., *London*
Prentice-Hall of Australia, Pty., Ltd., *Sydney*
Prentice-Hall of India, Pvt., Ltd., *New Delhi*
Prentice-Hall of Japan, Inc., *Tokyo*

Design by Jerrold J. Stefl, cover illustration by Tom Daly,
text illustrations by James Loates.

We wish that we had nothing to write under the title *Environmental Pollution*; but unfortunately that is not the case. Material is plentiful and is becoming more so every day.

We feel that, ultimately, survival of the human species depends upon the knowledge that people like you possess. Pollution has reached the stage where only a concerted effort by every citizen will reverse the current trend toward an ultimate "eco-catastrophe." To make this concerted effort, you need an understanding of the basic principles of ecology and a knowledge of the nature of the problems that threaten the biosphere. Therefore this is not a textbook to be studied for a test and then forgotten.

You will undoubtedly enjoy performing the investigations and experiments that are outlined in this book. The main function of the investigations and experiments, however, is not to entertain you. They are designed to increase your awareness of environmental problems and your comprehension of the causes and effects of environmental pollution. We hope that the ecological outlook which you develop by working with this book will help you to minimize your part in the degradation of the environment. Further, we hope that it will enable you to take your place with those citizens who are attempting to convince governments and industries that they, too, have a moral responsibility for the preservation of the environment. Reading this book won't help the environment; neither will performing the investigations. Only your action and the action of others like you can do that.

ACKNOWLEDGMENTS

This program was developed at the College of Education, University of Toronto. The resources of the College and the knowledge and skills of many student-teachers in the Environmental Studies option contributed greatly to the quality of the materials contained in this book and its companion volumes.

The authors are particularly appreciative of the competent professional help received from the Publisher. In particular, we wish to acknowledge the editorial assistance of Sue Barnes and Peter Anson. Their skill, knowledge, and patience are greatly appreciated. To Paul Hunt and Kelvin Kean we extend our sincere thanks for their help in planning this program in Environmental Studies. We are also grateful to Ron Decent and the many other members of the Prentice-Hall production staff for their effective work in the production of this book.

We wish also to thank Jim Loates for his excellent art work and Lois Andrews for her careful preparation of the manuscript.

W.A.A.
D.K.M.
A.L.R.

CONTENTS

6

FIELD AND LABORATORY STUDIES IN WATER POLLUTION   166

FIELD AND LABORATORY STUDIES IN AIR POLLUTION   202

7

CASE STUDIES   236

8

A GUIDE TO THE STUDY OF environmental pollution

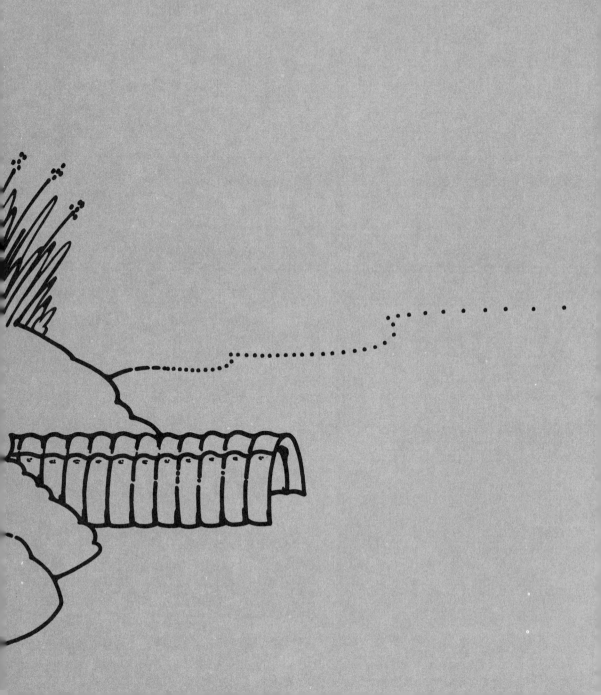

Basic Principles and Problems

1

In 1969 the citizens of Canada, the United States, and other "developed" countries woke up to the fact that the world is in an environmental crisis. In the past 300 years, man has wiped out at least 200 species of birds and mammals. In Canada alone, 66 species of wildlife are currently in danger of extinction, in all but one case due solely to human activity. The bald eagle (Fig. 1-1), peregrine falcon, Rocky Mountain bighorn sheep, wood bison, eastern cougar, and spotted turtle are on this macabre roll-call. Each day the list of species in danger of extinction grows longer. Is it possible that ultimately another name will be added to this list, that of man himself?

Fig. 1-1
How much longer will this species exist?

Fig. 1-2
This river looks clean but it smells like a sewer. Tests show that it carries a heavy load of human sewage.

Sewage, detergents, pesticides, and other chemical residues have changed sparkling rivers into foul-smelling sewers (Fig. 1-2). Clear blue lakes have become cesspools, unfit for most human needs. Ocean coastlines, once teeming with diverse plants and animals, are gradually turning into submarine deserts. Must we accept these changes as a necessary part of "progress"? What future changes do they forecast?

The earth receives about 70% of its life-sustaining oxygen as a by-product of the photosynthetic activity of marine plant life. Recently scientists discovered that DDT slows down photosynthesis in this plant life. In spite of this, many countries continue to use DDT to control insect pests because, they say, other pesticides are too expensive.

Mercury is toxic to humans in very low concentrations. In 1970, fishing was banned in portions of the Great Lakes system because some industries were dumping mercury residues into the water. Why did the industries do this? Should a plea of ignorance of the dangers of mercury free the industries from blame? Who should pay? Who is going to see that it never happens again? How many other toxic substances, as yet undetected, are knowingly added to our waterways?

Phosphates and nitrates from human sewage have contributed to algal blooms in many lakes, destroying their value as summer recreational areas. What will happen to man as a species if he is eventually denied recreational activities?

Sewage and industrial wastes have desecrated most of the waterways near large centers of population, yet sewage water can be purified enough to be recycled for drinking purposes. Why, then, do we continue to destroy our natural waters? How long can this destruction go on? It has been estimated that the expenditure of $2,000,000,000 (about four moonshots) by the U.S.A. and $300,000,000 by Ontario between 1970 and 1985 could clean up the Great Lakes and keep them that way. These costs seem small compared to expenditures for highways and for national defense. The cost per person in Ontario comes to about $1.70 per year, yet we are told that the high cost is the main reason that cleanup is progressing so slowly.

The ever-increasing human population is obviously at the heart of the pollution problem. What will ultimately happen if population growth goes on, unchecked? Can the necessary technology for maintaining suitable human conditions be developed fast enough to keep up with population growth?

The facts are clear. The crisis of the environment cannot wait another decade for answers. Only when each and every person demonstrates his concern through his actions as an individual and as a member of society will the crisis be met. The concern, however, must be an *informed* concern.

If you wish to conduct a meaningful study of any aspect of environmental pollution, you need a clear understanding of some basic ecological principles. In addition, you require knowledge of the specific environment to be studied and an overview of the problems that may confront that environment. This unit provides background for the aquatic environment. The discussion here is by no means exhaustive, and wide reading among the references is suggested. The more background you have, the more interesting and informative will be your field and laboratory studies.

1.1   WHAT IS POLLUTION?

In 1965 the Environmental Pollution Panel of the President's Science Advisory Committee, U.S.A., produced this definition of pollution:

> Environmental pollution is the unfavorable alteration of our surroundings, wholly or largely as a by-product of man's actions, through direct or indirect effects of changes in energy patterns, radiation levels, chemical and physical constitution and abundances of organisms. These changes may affect man directly, or through his supplies of water and. of agricultural and other biological products, his physical objects or possessions, or his opportunities for recreation and appreciation of nature.

This definition implies that pollution is not a problem for scientists only. Since it affects human lives, it is a health problem. Since it affects property and health, it is an economic problem. Since it affects living organisms, it is a problem in conservation of natural resources. Since it affects the senses, it is an aesthetic problem.

Man, like other organisms, has always polluted his environment with the by-products of his actions. As an organism, he creates wastes from his digestive and metabolic processes. As a social creature, he removes things from the environment and adds residues to it as he seeks housing, clothing, food, and relaxation for his family unit. So long as population density is low in a particular area, the environment is able to accommodate these alterations. When population density gets too high, however, deterioration of the natural environment—air, water, and soil—begins.

High population density is not the sole cause of the marked increase in pollution problems within the past two or three decades. Not only are populations growing in size, but they are also demanding a higher standard of living. Ever-increasing demands are made upon the earth's natural resources for industrial purposes which, in turn, result in the formation of wastes. Because of our industrial skill and genius, many of these wastes are virtually indestructible by natural forces. It seems that man has overpowered his environment with both his numbers and his creativity.

Why has this been allowed to happen? Many experts say that the basic reason is man's failure to recognize that he is an integral part of nature. Traditionally man has viewed himself as being in a constant struggle with nature. Over the years, society has rewarded the man who found new ways to exploit nature, and unrestricted economic growth has become of foremost importance to our society. Thus, to understand the present pollution problem and to take action that will halt and reverse the current trend, all of us need first to understand the interrelationships that exist between man and the rest of nature. This subject is the concern of a branch of biology called *ecology*, with emphasis placed upon the *ecosystem concept*.

For Thought and Research

Many sources define a *pollutant* as a substance that adversely affects something that *man* values, provided it is present in concentrations high enough to do so. What do you think of this definition? Discuss your opinion with others.

Recommended Readings

You can increase your knowledge of pollution problems by reading portions of any of these books:
1 *The Environmental Handbook* by Garrett de Bell, Ballantine Books, 1970.
2 *The Poison Makers* by R. D. Lawrence, Thomas Nelson & Sons, 1969.
3 *Moment in the Sun* by Robert & Leona Rienow, Ballantine Books, 1967.
4 *The Population Bomb* by Paul R. Ehrlich, Ballantine Books, 1968.
5 *Politics and Environment* by W. Anderson, Goodyear, 1970.
6 *The Environment* by the Editors of *Fortune*, Harper & Row, 1970.
7 *Pollution Probe* by D. A. Chant, New Press, 1970.
8 *Man in the Web of Life* by J. H. Storer, The New American Library, 1968.
9 *Eco-Catastrophe* by the Editors of *Ramparts*, Harper & Row, 1970.
10 *Man Against His Environment* by R. Rienow and L. T. Rienow, Ballantine Books, 1970.

1.2   THE ECOSYSTEM CONCEPT

Ecology deals with the interrelationships between living organisms and their environments. Ecologists, although commonly classified as biologists, generally find it necessary to move beyond the field of biology into geology, meteorology, physics, chemistry, and even into sociology and mathematics.

Ecologists find it convenient to recognize environmental units called ecological systems or ecosystems. An *ecosystem* is an interacting system consisting of groups of organisms together with their non-living or physical environment. The living organisms in an ecosystem—plants, animals, and protists—are referred to as the *biotic* components. The non-living portions, or *abiotic* components, include water, carbon dioxide, inorganic and organic substances in the soil, and such physical factors as wind, moisture, light, and temperature.

The most important thing about an ecosystem is that its various components are highly interrelated. Since each component, biotic or abiotic, is influenced by the others, the alteration of any one component will ultimately affect all of the others. Figure 1-3 illustrates this important fact. Study this figure carefully as follows: Pick a component such as "Soils." How may it influence those components to which arrows point? How may it be influenced by those components from which arrows point toward it? Do this for each component. Now think about the implications of these influences for environmental pollution.

A lake is an ecosystem, as is a forest, a field, or a classroom aquarium. At first glance these ecosystems appear to be

Fig. 1-3
Each component of an ecosystem, biotic or abiotic, is influenced by the other components. Can you explain the broken arrow between MAN and CLIMATE?

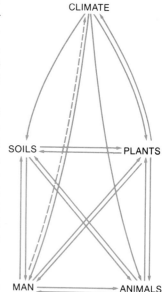

quite different; closer examination reveals that they share certain structural and functional characteristics. Let us see what they are.

Every ecosystem requires the input of energy. The only significant source is radiant energy from the sun. This energy is used by the *producers* of the ecosystem in the process of photosynthesis, in which carbon dioxide and water are chemically united to form energy-rich organic compounds such as glucose. The producers are mainly green plants, which contain the chlorophyll necessary for photosynthesis. Thus the algae of ponds, lakes, and oceans, the trees of forests, and the grasses of meadows are producers.

The organic compounds manufactured by the producers are used by them for their own growth and metabolism. Organisms that make their own food in this manner are called *autotrophic* (self feeding). The remaining organisms in an ecosystem are *heterotrophic* (other feeding). They must get their nutritional requirements by feeding on other organisms. Among the heterotrophic organisms are the *consumers. Primary consumers* (*herbivores*) feed directly on the producers. Cows, rabbits, and grasshoppers are primary consumers. *Secondary consumers* (*first-order carnivores*) feed on herbivores. Examples of first-order carnivores are cats, foxes, and insect-eating birds. They still derive their energy, though indirectly, from producers. Some ecosystems even have *tertiary consumers* (*second-order carnivores*) which feed on secondary consumers. Can you think of some examples?

The life forms in an ecosystem are all linked together through predator-prey relationships in what is called a *food chain*. All food chains follow this general pattern:

Producers → Herbivores → Carnivores → Higher Order Carnivores

In addition, all food chains terminate with a *top carnivore*, an animal which has no predators.

The study of food chains reveals that each organism, plant or animal, has a specific role in the ecosystem. Ecologists say that the organism occupies a *niche* in the ecosystem. The niche occupied by a green plant is that of producer, whereas a horse occupies the niche of herbivore. What niche(s) does man occupy in his food chain(s)?

It should be obvious that an organism can occupy a niche in more than one food chain. Thus it is impossible to draw parallel food chains to show what eats what in a particular ecosystem. Instead, these chains must be cross-linked. The resulting

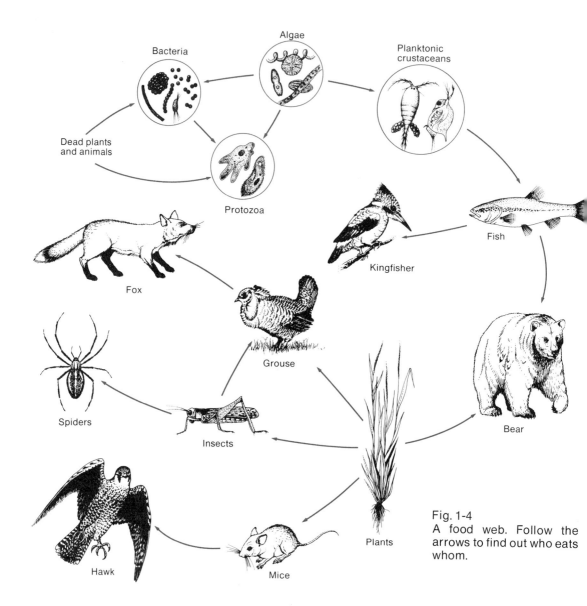

Fig. 1-4
A food web. Follow the arrows to find out who eats whom.

The images in the figure are labeled: Bacteria, Algae, Planktonic crustaceans, Dead plants and animals, Protozoa, Fish, Kingfisher, Fox, Grouse, Spiders, Insects, Bear, Hawk, Mice, Plants.

pattern is a *food web*. Most food webs have very complex patterns, as you can see from the example in Figure 1-4. The complexity increases when any of the organisms occupy more than one niche. Try to think of a plant, a terrestrial mammal, and an aquatic animal that occupy more than one niche in their respective ecosystems.

Ecological Pyramids. Since most predators are limited to prey of a rather narrow size range, food chains and food webs tend to proceed from very small organisms to progressively

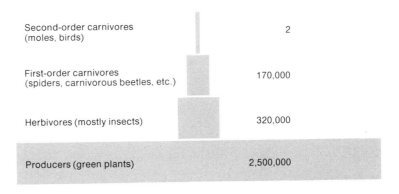

Fig. 1-5
A pyramid of numbers for a meadow. What do ecologists mean by "accumulation of pesticides in a food chain"?

larger organisms. Also, as the size of the organisms increases, the number of individuals decreases. This latter fact is often represented by a *pyramid of numbers* (Fig. 1-5). As you can see, this pyramid emphasizes the fact that a large number of organisms are required to support one organism at the next higher level in the food chain. Despite its simplicity, this pyramid is not considered very important. This is because it treats individuals of all species as though they were identical units. For example, a tulip plant, a rabbit, and a wolf each count as one unit. A 40-ton whale is equated numerically to a microscopic bacterium.

Fig. 1-6
A pyramid of biomass for a small, polluted lake. The numerals represent grams of dry biomass per square meter. In what ways would this pyramid be different if the lake were not polluted?

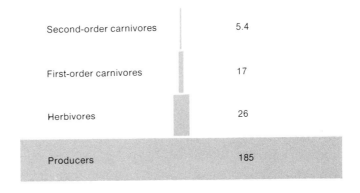

If you consider this matter carefully, you will see that a *pyramid of biomass* is of greater importance (Fig. 1-6). To construct this pyramid, one determines the total mass per unit area of each organism that is present in the food chain being studied. This type of pyramid implies that 1 gram of any organism provides the same amount of energy to a consumer as does 1 gram of any other organism. Yet this is not the case. For example, plants have, on the average, a caloric equivalent of about 4 kcal per gram. (This means that complete burning of 1 gram of the substance releases 4 kcal of heat energy.) Animals, on the other hand, have an average caloric equivalent of about 5 kcal per gram. Thus you can obtain more energy by eating 1 gram of animal than you can by eating 1 gram of plant. Scientists have discovered that the tissues of seeds, migrating birds, and hibernating animals have caloric equivalents as high as 7 or 8 kcal per gram. Why is this so?

Of more fundamental importance is the *pyramid of energy* (Fig. 1-7). Each level in this pyramid represents the total energy flow at that level. This includes the energy tied up in the formation of new tissue and that released by respiration. The data required for such a pyramid are very difficult to determine. Yet, when the pyramid is constructed, large and small organisms are put into true perspective. The efficiency with which energy is captured and passed on to the next level is more important than the numbers or sizes of the organisms in the food chain.

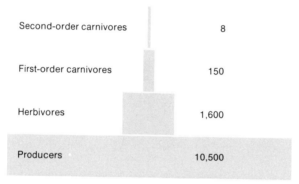

Second-order carnivores	8
First-order carnivores	150
Herbivores	1,600
Producers	10,500

Fig. 1-7
A pyramid of energy for an aquatic ecosystem. Each figure represents the total energy flow at that level (kilocalories per square meter per year).

Energy Flow in Ecosystems. Two factors concerning energy flow in ecosystems should be apparent to you. First, energy is lost to the environment at each successive level. Why is this so? Second, the flow of energy through an ecosystem is one-way; energy is not cycled within the ecosystem. In other words, herbivores or carnivores cannot provide producers with the energy they require for life. This means that energy must constantly enter every ecosystem from without, generally from sunlight.

More detailed discussion of energy flow in ecosystems can be found in *Recommended Readings* 1 and 5.

Nutrient Cycles in Ecosystems. Although an ecosystem cannot function without the input of energy, this input of energy, alone, cannot cause an ecosystem to function. Over 20 different elements must be present to sustain life processes in an ecosystem. The chief elements required are carbon, hydrogen, oxygen, nitrogen, phosphorus, and sulfur.

Unlike energy flow, the flow of these elements is cyclic within an ecosystem; thus there need not be a constant input of the elements. Producers pass the elements on to herbivores, and herbivores pass them on to first-order carnivores. And so the elements pass along the food chain. How do they get back to the producers to complete the cycle? An important group of organisms occupying the niche of *decomposers* perform this task. Decomposers are largely bacteria, yeasts, and molds. They break down animal excretions and dead organisms into simpler components that can be taken in and reused by producers (green plants). Frequently microorganisms called *transformers* must act on the materials formed by the decomposers before these materials can be used by plants. Specific examples of the roles of decomposers and transformers are given in Unit 2. Ecologists refer to these cycles of elements as *nutrient cycles* or *biogeochemical cycles*.

Three non-essential niches are commonly found in most ecosystems. They need not be present for the ecosystem to function, but they often facilitate its functioning. *Parasites*, which may be carnivorous or herbivorous, obtain energy from living hosts without actually killing them. Most species of plants and animals are hosts for parasites at some time or other. Can you think of some examples? Is man host to any parasites? The niche of *scavenger* is occupied by animals such as crows, vultures, ants, and termites. These animals eat dead plant or animal matter, or, in some cases, both. In doing so, they break down the dead matter to the point where decomposers and transformers can more readily act on it. Finally, *saprophytes* constitute a special group of scavengers. This niche is filled mainly by fungi, such as mushrooms and molds. Being heterotrophic, these plants obtain their energy by absorbing organic matter from dead things such as logs, stumps, and the humus of soil.

Summary of the Structure and Functioning of Ecosystems. You must have a clear understanding of the ecosystem concept before you can meaningfully study environmental pollution. Do you feel that you now have that understanding? Test yourself by reading this summary of key ideas. If you do not understand any of them, reread Section 1.2 and, if necessary, consult the appropriate *Recommended Readings*.

1)   Although ecosystems may differ widely in species composition, they all require the same three biological com-

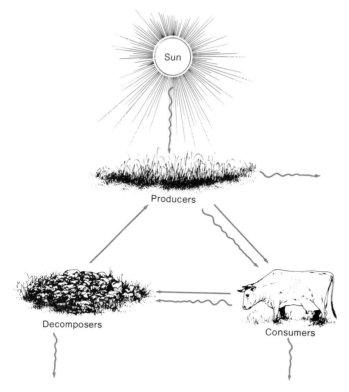

Fig. 1-8
A simplified model of the structure of an ecosystem, showing the cyclic flow of nutrients and the non-cyclic flow of energy in the system.

ponents: producers, consumers, and decomposers (Fig. 1-8).

2)	Energy flow in ecosystems is one-way. Energy is progressively lost along the food chains in ecosystems. Thus an outside source of energy is required. (Follow the wavy brown arrows in Figure 1-8).
3)	All ecosystems require essentially the same basic nutrients. These nutrients are recycled within each ecosystem. (Follow the straight brown arrows in Figure 1-8.)
4)	A highly interdependent relationship exists among all components, biotic and abiotic, of an ecosystem (Fig. 1-3).

For Thought and Research

1 Protists were omitted from Figure 1-3 in order to avoid overwhelming you with a maze of arrows. Try drawing an ecosystem that includes protists. You might also draw one in which "Climate" is replaced by two terms, "Macroclimate" and "Microclimate."

2 Complete these food chains:
(a) In a prairie grassland: grass → grasshoppers → . . .
(b) In Lake Ontario: microscopic plants → microscopic animals → . . .

(c) In a temperate deciduous forest: herbaceous plants → plant lice → . . .
Compare your results with those of others and draw appropriate conclusions.

3 Be sure that you clearly understand the terms "food chain" and "food web." Why is this understanding an important prerequisite to the study of environmental pollution? If you wish to read further on these topics, you will find an interesting and informative account in *Recommended Reading* 2.

4 Pyramids of energy always exhibit the same general shape as the one shown in Figure 1-7; that is, these pyramids become progressively more narrow as one moves to higher trophic levels. Why? Occasionally a pyramid of numbers or a pyramid of biomass may be inverted. Why would this be so? You will find this question easier to answer if you first try to think of two or three examples of inverted pyramids. For example, under what circumstances would herbivores outnumber producers in an ecosystem?

5 Consider the nature of energy flow in ecosystems. What message does this convey to a world with an ever-expanding population to be fed?

6 You are probably familiar with the water cycle. It plays an important role in the functioning of natural ecosystems, yet it is not commonly classified as a biogeochemical cycle. Why?

7 When studying environmental pollution why is it important to be aware of the fact that nutrients such as phosphorus and nitrogen are recycled in ecosystems?

8 The model of an ecosystem in Figure 1-8 is a simplified one. Try the following:

(a) Enlarge the model by dividing consumers into herbivores and carnivores. Compare your diagram with the one on page 4 of *Recommended Reading* 5.

(b) Expand the model to show man's role as a harvester and manipulator. In this role man seeks to increase the flow of energy and nutrients to himself. Compare your diagram with the one on page 26 of *Recommended Reading* 1.

(c) Expand the original model to include every niche that we have mentioned in this section. Compare your diagram with the one on page 467 of *Recommended Reading* 6.

Recommended Readings

1 *Readings in Conservation Ecology* by G. W. Cox, Appleton-Century-Crofts, 1969.
2 *Basic Ecology* by Ralph and Mildred Buchsbaum, Boxwood Press, 1957.
3 *Ecology* by E. P. Odum, Holt, Rinehart & Winston, 1963.
4 *Ecology* by Peter Farb, Life Nature Library, Time, Inc., 1963.
5 *Concepts of Ecology* by E. J. Kormondy, Prentice-Hall, 1969.
6 *Elements of Ecology* by G. L. Clarke, John Wiley & Sons, 1966.
7 *Scientific American*, September, 1970. This volume is devoted entirely to papers on energy flow and nutrient cycling in the biosphere.

1.3 CONSTRUCTING A CLASSROOM ECOSYSTEM

This is a long-term experiment that will permit you to study carefully the many aspects of an ecosystem introduced in Section 1.2. You need a large jar or bottle, with a capacity of at least 2

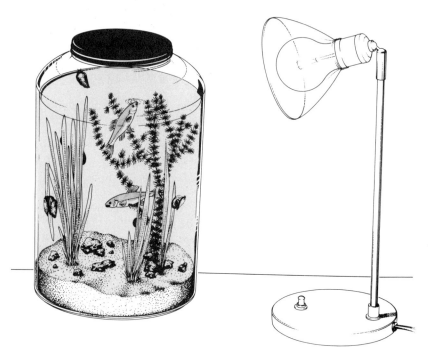

Fig. 1-9
A classroom ecosystem.

or 3 gallons. You must be able to stopper the bottle to eliminate gas exchange between the atmosphere and the ecosystem.

Place an inch or two of sand and gravel in the bottom; fill the bottle with water. If you use tap water, let the water dechlorinate for 48 hours or more. Add a few strands of an aquatic plant such as *Elodea*. You can encourage the *Elodea* to root in the bottom by tying a small stone to the base of each strand. A few spoonfuls of a small floating aquatic plant such as duckweed can also be added. Now place a few snails and one or two small fish (such as guppies) in the water. Stopper the bottle. Place one or two 100-watt light bulbs in the position shown in Figure 1-9.

Observe your ecosystem for almost a full year. During the first few weeks, you may have to change the positions of the lights and add or remove plants and animals until a balance is attained. If any of the animals die, the nutrients released by their decay may cause an algal bloom (identified by a green color in the water). If you wish to eliminate the unsightly algae, siphon off some of the nutrient-rich water or, better, add animals to eat the algae. Snails and small catfish eat dead algae. Your local aquarium store can recommend fish that eat living algae.

One factor should be kept in mind when making your observations. Your ecosystem is a *closed* ecosystem. How does this differ from natural ecosystems? What advantages are gained by studying a closed ecosystem? What niche does each of the living organisms, plant and animal, occupy in this ecosystem? What nutrient cycles are taking place? What niches are occupied by microorganisms? Account for any changes that occur in the ecosystem as time progresses. What evidence do you have that energy flow is uni-directional in ecosystems?

1.4 PHOTOSYNTHESIS AND RESPIRATION— MAINTAINING A LIFE BALANCE

You probably have studied photosynthesis and respiration before. You may not, however, have had an opportunity to perform experiments on these topics. Unit 6 contains a number of experiments concerning photosynthesis and respiration or, if you like, concerning the roles of producers, consumers, and decomposers in ecosystems. After you read this section, you should perform those experiments that you have not successfully completed at some prior time.

Imagine, for a moment, that your home community, with a few square miles of the surrounding countryside, is enclosed in a gigantic, transparent plastic dome (Fig. 1-10). This closed system resembles the classroom ecosystem in Section 1.3. Can this closed ecosystem operate normally for a long period of time, say for several months? We can answer this question by looking at the requirements of ecosystems. First, this ecosystem appears to have the 3 essential biological components. It has producers— the green plants of the woods, fields, marshes, and ponds; it has consumers—cows, dogs, insects, and people; it has decomposers—bacteria, yeasts, and molds. Second, this ecosystem has

Fig. 1-10
Will this ecosystem continue to function normally?

a transparent dome through which sunlight can enter. Finally, an ecosystem this large will likely contain the nutrients that are essential for life. Since nutrients are recycled, enclosing the ecosystem should create no problems. It appears as though the ecosystem *could* continue to function. Maybe this is an answer to environmental pollution—just enclose your community in a plastic dome.

There is a problem, though, and it's a big one. Before we put the dome in place, the ecosystem was an open one. As such, it received materials from neighboring ecosystems. You can undoubtedly think of many things that came into your community before it was enclosed. Cars, other manufactured goods, gasoline, oil, food, and textiles are obvious things. Not so obvious, but most important of all, is oxygen. Did you know that, if your community is an average North American community, about 60–70% of the oxygen which it requires comes from the oceans? If you live in a highly industrialized community, the percentage is much higher than that. Now you know why ecologists warn us about the long-term dangers of water pollution. Some people are prepared to tolerate rivers that are open sewers and lakes that are cesspools because "that's part of modern living." They forget, though, that the contaminated water of these rivers and lakes eventually ends up in the oceans, the same oceans whose phytoplankton (microscopic plants) produce the bulk of our oxygen supply. Indeed, there is already some evidence that pollutants entering the oceans are affecting the phytoplankton.

By now you have probably concluded that our imaginary closed ecosystem could not continue to function. Well, don't be too hasty. Is it possible that, through some readjustment in its biological community, the ecosystem could continue to function? Suppose, for example, that those organisms which consume large amounts of oxygen and which demand unusual things like gasoline were to die. Could the remaining organisms, with their physical environment, form a new functioning ecosystem? Could the producers present form enough oxygen to meet their own respiratory needs as well as those of a limited number of consumers and decomposers? Could all the living organisms present produce enough carbon dioxide for the photosynthetic requirements of the producers? Let's first look into photosynthesis and respiration a little further. The following discussion is brief, since detailed treatment of these topics is given in most introductory biology books.

Respiration. The cells of all living things require a constant supply of energy in order to carry out essential cell processes such as growth, repair, and reproduction. Where do they get this energy? They get it by oxidizing or "burning" a fuel, just as you get energy to heat your home by burning a fuel. The fuels used by cells and those used in your home are more similar than you might expect. Both are organic in nature; that is, they consist of compounds containing carbon atoms that were once associated with living organisms. When you burn wood, coal, or fuel oil, the carbon atoms in the organic compounds are separated from one another, due, in large part, to the oxygen gas that must be present. As the atoms separate, the chemical energy of the bonds that held them together is released, generally in the form of heat and light. Carbon dioxide and water are formed as by-products.

The "burning" of fuels within cells is called *cellular respiration*. The principal fuel for this oxidation reaction is *glucose*, a simple sugar found in many foods. You can prove to yourself that glucose is an organic compound by heating a small amount of it in a test tube. Try it! Glucose, in the presence of oxygen, burns readily when heated. To permit "burning" at temperatures that will not be damaging, cells have *catalysts*. They lower the temperature at which glucose oxidizes. Organic catalysts which operate in living cells are called *enzymes*. When glucose is oxidized in living cells, energy is released and carbon dioxide and water are formed as by-products. The following word equation summarizes the process of cellular respiration:

$$\text{Glucose} + \text{Oxygen} \xrightarrow{\text{enzymes}} \text{Energy} + \text{Carbon Dioxide} + \text{Water}$$

The chemical equation is:

$$C_6H_{12}O_6 + 6\,O_2 \xrightarrow{\text{enzymes}} \text{Energy} + 6\,CO_2 + 6\,H_2O$$

The energy released is used by cells for their life processes.

Photosynthesis. Where do cells get the glucose in the first place? That depends on the type of cell. If the cell is an autotroph (producer), it makes its own glucose. A cell of this type contains chloroplasts. They, in turn, contain the chlorophylls that help the cell to convert light energy into stored chemical energy. How is this done? You read earlier that energy is released when large molecules are broken up into smaller ones. It seems reasonable to expect, then, that energy is required when smaller molecules are combined to form larger molecules. That is what happens during *photosynthesis*. As the name implies, light energy (photo) is used to build (synthesize) complex substances out of simpler substances. This process occurs only in green plants.

The cells of these plants take in carbon dioxide and water and, with the assistance of chlorophyll and light energy, synthesize organic compounds like glucose. In addition, a very important by-product is formed, oxygen gas. During photosynthesis, then, light energy is converted to and stored as chemical energy in the bonds between the carbon atoms in glucose molecules. This energy can be released at some later time by oxidation of the glucose. The word equation and the chemical equation for photosynthesis are:

Carbon Dioxide + Water + Light Energy
$$\xrightarrow{\text{chlorophyll}} \text{Glucose} + \text{Oxygen}$$

$6 CO_2 + 6 H_2O + \text{Light Energy}$
$$\xrightarrow{\text{chlorophyll}} C_6H_{12}O_6 + 6 O_2$$

Both photosynthesis and respiration are much more complex than these equations suggest. These are merely summation equations that indicate the starting materials and the final products. Other complex reactions take place between these stages.

Obviously not all cells can photosynthesize. If a cell is in a heterotroph, it cannot make its own glucose and, as a result, must get it from its environment. This is why herbivores eat producers and why first-order carnivores eat herbivores. Although the predator may not ingest glucose directly when it eats its prey, its digestive enzymes soon form glucose out of many of the food materials that it does ingest.

This completes our brief story of photosynthesis and respiration. As you reconsider the plastic-domed ecosystem, think about the roles of producers, consumers, and decomposers in the maintenance of an oxygen-carbon dioxide balance. This balance is essential for life as we know it in our ecosystem, the biosphere.

For Thought and Research

1 Consider these questions:
(a) What would happen if the oxygen concentration of the atmosphere increased? Decreased?
(b) What would happen if the carbon dioxide concentration of the atmosphere increased? Decreased?
(c) The oxygen and carbon dioxide content of a river or lake is governed to some extent by the plants and animals in the body of water. The concentrations of these gases in the water are also regulated by a number of physical factors in and

around the body of water. What are these factors and in what ways do they affect the concentrations of the two gases?

2 Perform the necessary experiments in Unit 6.

Recommended Readings

For further information on photosynthesis and respiration consult:

1 *Biological Science: Molecules to Man*, B.S.C.S. Blue Version, Houghton Mifflin, 1969.

2 *Biological Science: An Inquiry into Life*, B.S.C.S. Yellow Version, Harcourt Brace Jovanovich, 1969.

3 *High School Biology*, B.S.C.S. Green Version, Rand McNally, 1969.

4 *The Spectrum of Life* by H. A. Moore and John R. Carlock, Harper & Row, 1970.

5 *Modern Biology* by J. H. Otto and A. Towle, Holt, Rinehart & Winston, 1968.

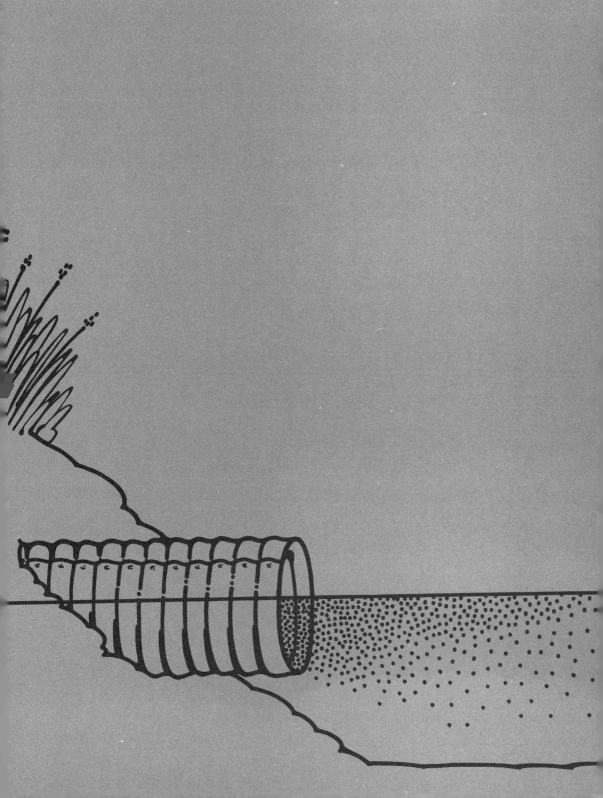

Chemical and Physical Aspects of Water Pollution

2

One of your main objectives during a field trip to a polluted body of water is to collect data on the physical and chemical properties of the water. Another objective is to determine the species of living organisms present and the abundance of each species. Then back in the laboratory you perform the most challenging and useful part of your study—to relate the physical and chemical data to the biological data that you have collected. In other words, you try to determine how the water quality affects the life in the water. This is not easy since many interrelated factors are involved.

You must be sure that you do not draw conclusions that fail to take into account some of these factors. For example, each animal species has a temperature limit above which it cannot live. Yet, for a particular species, no *one* upper limit can be stated because this value depends on many other factors. For instance, lobsters in 30% salt solution have a lethal temperature of 32°C when the solution has a dissolved oxygen content of 6.4 mg per liter; their lethal temperature in the same solution drops to 29°C when the dissolved oxygen content is lowered to 2.9 mg per liter. (Why would this be so?) It would therefore be wrong to state that raising the temperature increases the lobster death rate unless you specified that other environmental factors were kept constant. This means, of course, that you must measure and consider as many factors as you possibly can during a pollution study. You must be familiar with the physical and chemical factors that are to be measured and with their significance. Unit 2 will help you gain this familiarity.

Perhaps you would like to see how much you already know before you read on. Table 1 gives representative data for the physical and chemical properties of a body of water. Study it carefully and see if you understand it. If you don't, read this unit and then try again. Do not proceed with water pollution studies until this table is meaningful to you.

TABLE 1 A SAMPLE WATER ANALYSIS

Temperature	15.2°C
Total suspended solids (T.S.S.)	61 ppm
Total dissolved solids (T.D.S.)	225 ppm
Dissolved oxygen (D.O.)	6.5 ppm
pH	7.8
Biochemical oxygen demand (B.O.D.)	6.0 ppm
Chemical oxygen demand (C.O.D.)	11.0 ppm
Alkalinity	60 ppm
Hardness	55 ppm
Color	6 units
Phosphates	0.9 ppm
Nitrates	1.7 ppm
Ammonia	0.8 ppm
Coliforms	70 per 100 ml
Secchi disc reading	4.5 meters
Carbon dioxide (free)	0.75 ppm

2.1 DISSOLVED OXYGEN (D.O.)

Oxygen is a clear, colorless, odorless, tasteless gas that dissolves to a limited extent in water. Aquatic organisms, both plant and animal, depend on dissolved oxygen for survival. Thus the determination of the dissolved oxygen content is probably the most significant test you can perform to measure the quality of the water in a stream, pond, or lake (Fig. 2-1). In general, an acceptable environment for aquatic life must contain no less than 5 ppm of oxygen (5 mg of oxygen per liter of water). Variations from this figure are wide, of course, and depend on the nature of the organism, its degree of activity, the pollutants present, the water temperature, and other factors. Here are some examples of these variations.

The trout species *Salmo trutta* requires, under normal conditions, oxygen concentrations in excess of 10 ppm. The chub, *Leuciscus cephalus*, requires only 7 ppm. The carp, *Cyprinus carpio*, can remain alive in water containing as little as 1–2 ppm of dissolved oxygen. Can you account for these different oxygen requirements?

One species of trout uses about 55 mg of oxygen per hour for each kg of body weight when it is resting at a temperature of 5°C. At 25°C, under similar conditions, the consumption rises to approximately 285 mg per hour. In contrast, one species of goldfish has an oxygen consumption of only 14 mg per hour per kg of body weight when it is resting at 5°C, but the consumption increases ten-fold to 140 mg per hour when the temperature is raised to 25°C. Obviously the rate of oxygen consumption by fish depends on both temperature and species.

Solubility of Oxygen in Water. It is not a coincidence that fish, such as trout, with high oxygen requirements are found in cold water; nor that fish like carp that can tolerate low-oxygen conditions are found in warm water. The selection of a habitat by these fish and by other aquatic organisms is regulated to a large extent by the fact that the solubility of oxygen, like that of all gases, decreases as the temperature increases.

Table 2 gives the solubility of oxygen gas in water when air (21% oxygen) is in contact with the water at standard pressure (760 mm). Graph these data and study the results carefully.

Before you go on a field trip, you should master the techniques involved in a D.O. determination. Check yourself by confirming one or two of the values in this table. Use a Hach or LaMotte kit, or the Winkler method (Section 6.7). Let the water stand at the chosen temperature for several hours with slow aeration before doing the D.O. test. An aquarium heater will help you maintain the required temperature. Your value should agree with ours to ±0.5 ppm. Retain the graph; it will be of value to you during pond and stream studies.

Oxygen as a Limiting Factor. In lakes oxygen can often be the limiting factor in determining the life-forms present. Lakes get much of their oxygen from the atmosphere; the remainder comes from photosynthesizing organisms within the

Fig. 2-1
Pond water being tested for its dissolved oxygen content.

TABLE 2   SOLUBILITY OF OXYGEN IN WATER
(When air containing 21% O_2 is in contact with
water at 760 mm pressure)

Temperature (°C)	Solubility (ppm)
0	14.6
5	12.7
10	11.3
15	10.1
20	9.1
25	8.3
30	7.5

lake. If oxygen is consumed by the respiration of organisms within the lake faster than it is replenished, an oxygen deficit results. This prohibits the existence of life-forms that otherwise could live there. At the moment, Lakes Superior and Huron are saturated with oxygen to all depths. Lake Michigan is close to saturation at all depths, but is sometimes only 70–80% saturated in deep areas. Lake Ontario, however, is only 50–60% saturated in deep water; the central basin of Lake Erie is below 10% saturation in its deeper regions.

Oxygen is consumed in the bottoms of these latter lakes by the many decomposers feeding upon sewage, dead algae, and the like. It is not replaced rapidly enough to prevent a deficit because this deep water is below the zone of light penetration. Hence no photosynthesis occurs. Also, the movement of oxygen from the surface to the bottom requires a long time interval. Thus, although the bottom of Lake Erie may be a suitable habitat for trout with regard to food supply and temperature, the low oxygen content is the limiting factor that prevents trout from living there.

Dissolved oxygen is generally not a limiting factor in rapidly flowing, clean rivers. In such rivers the concentration of oxygen usually stays at the saturation point. Why is this so? In sluggish or polluted streams, however, the D.O. level may be well below saturation and, therefore, a limiting factor.

In summary, the D.O. concentration of a body of water depends on:

1)	the temperature;
2)	the presence or absence of photosynthetic plants (microscopic and macroscopic);

3)   the degree of light penetration (dependent upon depth and turbidity);

4)   the degree of turbulence in the water;

5)   the amount of organic matter being decomposed in the water (sewage, dead algae, and industrial wastes).

Be sure that you understand clearly how each of these factors affects the dissolved oxygen content of the water. Consult the *Recommended Readings* for further assistance.

A single D.O. determination in a river or lake is of little value if you wish to assess the general condition of the body of water. You should make several determinations at different depths, locations, and times of day. For example, consider a river that has sewage entering it. The D.O. level above the sewage plant might be 8.0 ppm at a given time; 500 yards below the sewage outlet the D.O. level might be only 2.1 ppm. One mile further downstream, it could be back up to 8.0 ppm again. Further, these values could change dramatically during flood conditions or periods of drought.

If the D.O. concentration is consistently low in a body of water, not only is desirable life affected, but *anaerobic* organisms (those not requiring oxygen) begin to decompose the organic matter in the water, liberating by-products such as hydrogen sulfide and methane. These gases are responsible for the stench of many stagnant and polluted bodies of water.

For Thought and Research

If you plan to carry out pollution studies in a reasonably large lake, you need to know more about the physical nature of lakes than is outlined here. Consult *Recommended Readings* 1, 3, 4, and 5. You should clearly understand the meanings of the following terms: *thermal stratification, epilimnion, thermocline, hypolimnion, spring overturn, fall overturn, homothermous, tropholytic zone, trophogenic zone, compensation depth, littoral zone,* and *limnetic zone.*

Recommended Readings

1   *A Guide to the Study of Freshwater Ecology* by W. A. Andrews et al., Prentice-Hall, 1972.
2   *The Biological Aspects of Water Pollution* by C. G. Wilber, C. C. Thomas, 1969.
3   *Fundamentals of Limnology* by F. Ruttner, University of Toronto Press, 1963.
4   *The New Field Book of Freshwater Life* by E. B. Klots, Putnam, 1966.
5   *Streams, Lakes, Ponds* by R. E. Coker, Harper & Row, 1968.

2.2 CARBON DIOXIDE IN WATER

The dissolved oxygen concentration in a body of water is, in many ways, dependent upon the concentration of carbon dioxide in the water. Therefore, it is important to consider in some detail the factors that affect the concentration of carbon dioxide in water. Let us begin by studying the solubility of carbon dioxide gas in water when air (0.03% carbon dioxide) is in contact with water at standard pressure. Graph the data in Table 3. Compare the results with those that you obtained for oxygen gas.

TABLE 3 SOLUBILITY OF CARBON DIOXIDE IN WATER
(When air containing 0.03% CO_2 is in contact with water at 760 mm pressure)

Temperature (°C)	Solubility (ppm)
0	1.00
5	0.83
10	0.70
15	0.59
20	0.51
25	0.43
30	0.38

For which gas does the solubility exhibit the greater temperature dependence? What significance do you attach to this fact?

Carbon Dioxide from the Atmosphere. Carbon dioxide, like oxygen and other gases, is continually exchanged between the atmosphere and water that is in contact with the atmosphere. If a body of water is calm, one would expect this exchange to occur only through the process of diffusion. This process is extremely slow under most conditions. It has been estimated, for example, that it would require up to a million years for carbon dioxide and oxygen from the atmosphere to saturate water in a lake to a depth of 50 meters, if diffusion were the only means by which the gases could travel through the water. Yet experimental evidence has shown that deep waters can undergo a marked change in dissolved oxygen or carbon dioxide within a few days. Such a rapid change cannot be explained by diffusion. Obviously wave action helps the exchange of gases at the surface, but the movement of gases into and out of deep regions depends upon another process. In this remarkable phenomenon, called *spring* *overturn* or *fall overturn*, large masses of water circulate from the top of a lake to the bottom, carrying with them dissolved gases. See *For Thought and Research* page 26.

Carbon Dioxide in Rain Water. Carbon dioxide also enters natural waters in rain water. As raindrops fall, they absorb carbon dioxide from the air. Normally the quantity absorbed does not exceed 0.6 ppm. Once the carbon dioxide is dissolved in water, most of it does not remain in the form of molecules of carbon dioxide (CO_2). Instead, it reacts with the water to form a weak acid called carbonic acid:

$$CO_2 + H_2O \rightleftarrows H_2CO_3$$

This acid is present in the soda water used to make most soft drinks. Why does "pop" fizz when the top is removed from the bottle?

The carbonic acid, in turn, partially dissociates (separates) into hydrogen ions and bicarbonate ions:

$$H_2CO_3 \rightleftarrows H^+ + HCO_3^-$$

The bicarbonate ions also decompose to a certain extent:

$$HCO_3^- \rightleftarrows H^+ + CO_3^{2-}$$

The hydrogen ions from these two reactions are responsible for the acidic properties of a carbon dioxide solution. A sour taste and the ability to turn litmus paper red are two such properties.

The three reactions that take place when carbon dioxide dissolves in water can be summarized as follows:

$$CO_2 + H_2O \rightleftarrows H_2CO_3 \rightleftarrows H^+ + HCO_3^- \rightleftarrows H^+ + CO_3^{2-}$$

Do you see the significance of the double arrows?

The amount of carbon dioxide that is present in the solution as CO_2 and H_2CO_3 is referred to as the *free carbon dioxide*. The carbon dioxide present as HCO_3^- and CO_3^{2-} is called the *combined carbon dioxide*. The significance of the free and combined forms will be considered later.

Now, let us consider rain water falling on land, rather than directly into water. As it moves down through soil, it comes in contact with still more carbon dioxide, largely in the air spaces between the soil particles. Finally the rain water, now a fairly concentrated solution of carbonic acid, comes in contact with calcium-bearing rocks like limestone. A reaction occurs which forms calcium bicarbonate:

$$CaCO_3 + H_2CO_3 \rightarrow Ca(HCO_3)_2$$

Unless excess carbon dioxide is present, this calcium bicarbonate decomposes readily:

$$Ca(HCO_3)_2 \rightarrow CaCO_3 + H_2O + CO_2$$

The calcium carbonate ($CaCO_3$) formed in this reaction precipitates out as limestone. If excess carbon dioxide is present (as is commonly the case in soil), the calcium bicarbonate remains stable and eventually ends up in a body of water as a result of runoff.

In summary, much of the carbon dioxide that enters natural waters ends up as a carbonate or bicarbonate of calcium. If the soil contains magnesium-bearing rocks like dolomite, some of the carbon dioxide will end up as a carbonate or bicarbonate of magnesium. Most natural waters contain all of these compounds. Neither bicarbonate can exist for long, however, without *free carbon dioxide* in the water. Surface water usually contains less than 10 ppm of free carbon dioxide. Water that contains over 25 ppm of free carbon dioxide can be lethal to many organisms.

Carbon Dioxide from Metabolism. Other factors that affect the carbon dioxide content of natural waters are photosynthesis and respiration. You should be able to figure this out by starting with a few questions:

1) Photosynthesis consumes carbon dioxide. How might this fact affect the carbon dioxide concentration
(a) at the bottom of a very deep lake?
(b) at the margin of the same lake?
(c) at the surface of a lake or pond?
(d) 2 feet down in an old pond?

2) Respiration by living organisms, plant and animal, releases carbon dioxide. How will this affect the carbon dioxide concentration at each of the sites above during daylight periods? At night?

3) Consider carefully the action of *aerobic* decomposer organisms as they break down organic residues on the bottoms of ponds and lakes. What effect will they have on the carbon dioxide concentration?

4)   Changes in the carbon dioxide concentration are usually accompanied by changes in the dissolved oxygen concentration. For each of the questions above, determine what changes in D.O. concentration accompany the changes in carbon dioxide concentration.

The Carbon Cycle. By now it should be apparent that carbon is one of those elements that is cycled through ecosystems. (Recall Section 1.2.) The most obvious part of this cycle involves the movement of carbon from the atmosphere through producers, consumers, and decomposers, and then back to the atmosphere again. The carbon in the atmosphere is, of course, in the form of carbon dioxide. It is fixed in energy-rich compounds like glucose by the producers, which then pass it on to the consumers. Finally it is returned to the atmosphere by the decomposers that feed on dead producers and consumers. Along this pathway, however, some carbon is returned to the atmosphere as the producers and consumers respire.

A similar pathway is followed by carbon atoms in aquatic ecosystems. The overall carbon cycle is made more complex by the constant exchange of carbon dioxide between the air and the water. It is important to note that water, largely the oceans, holds over 50 times as much carbon dioxide as the air. Therefore it tends to regulate the carbon dioxide in the air.

All parts of this cycle are interdependent, as shown in Figure 2-2. Examine this diagram carefully to see the interdependence. Pay particular attention to how the cycle adjusts to additions and subtractions of carbon at various points in it.

Fig. 2-2
The carbon cycle.

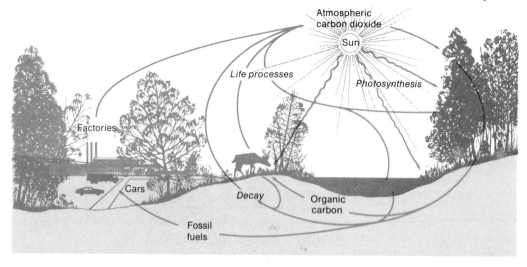

For Thought and Research

1 Use the carbon cycle to trace the chain of events that is likely now occurring within the biosphere as a result of the addition of carbon dioxide to the atmosphere through the combustion of fossil fuels.

2 How would an aquatic ecosystem respond to the sudden death of a high percentage of its producers?

3 The addition of particulate matter to the air from such sources as smoke stacks and automobile exhaust is gradually reducing the amount of sunlight reaching the surface of the earth. Predict, through reference to the carbon cycle, the long-term effects of the continued addition of such matter to the air.

Recommended Readings

1 *Streams, Lakes, Ponds* by R. E. Coker, Harper & Row, 1968.
2 *Fundamentals of Limnology* by F. Ruttner, University of Toronto Press, 1963.
   For further information on the carbon cycle, consult:
3 *Concepts of Ecology* by E. J. Kormondy, Prentice-Hall, 1969.
4 *Ecology and Field Biology* by R. L. Smith, Harper & Row, 1966.
   The most thorough discussion of carbon dioxide in the aquatic environment is found in:
5 *Elements of Ecology* by G. L. Clarke, John Wiley & Sons, 1966.

Fig. 2-3
The *p*H scale.

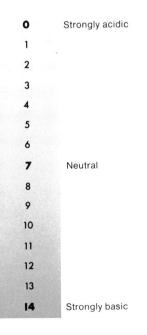

0	Strongly acidic
1	
2	
3	
4	
5	
6	
7	Neutral
8	
9	
10	
11	
12	
13	
14	Strongly basic

2.3   *p*H, ACIDITY, ALKALINITY, AND HARDNESS

You have seen that a close relationship exists between the oxygen and carbon dioxide content of natural waters. When one of these factors changes, the other usually changes. If the relationship were as simple as we have implied, solving many of our pollution problems would be a relatively easy task. Unfortunately, many other factors are also related to the carbon dioxide content and, as a result, indirectly affect the dissolved oxygen content of water. Four such factors are the *p*H, acidity, alkalinity, and hardness of the water. Since these four factors are highly interrelated, they are best considered together.

   (a) pH. The *p*H of an aqueous solution represents the concentration of hydrogen ions in the solution. It is a scale that runs from 0 to 14. On this scale, 7 is neutral, below 7 is acidic and above 7 is basic. In other words, the acidity of a solution decreases and its basicity increases as the *p*H goes from 0 to 14 (Fig. 2-3). The chemistry behind the value of *p*H does not concern us here. (If you wish to explore it further, see *Recommended Readings* 1, 2, and 3.) Our prime concerns here are the *limiting* nature of *p*H and its relationship to other environmental factors.

Recent studies have shown that waters having a *p*H range from 6.7 to 8.6 will generally support a good fish population. So long as the *p*H is within this range, it appears to have no effect on life processes like growth and reproduction. In fact, most fish species can tolerate *p*H values beyond this range. Only a few species, however, can tolerate *p*H values lower than 5 or greater than 9. Those species are usually so specifically adapted to the condition of high or low *p*H that they cannot tolerate *p*H values in the normal range.

The *p*H of a body of water generally drops as the body of water ages (see Section 3.2). A body of water is usually basic when young and it becomes more acidic with time. This is caused chiefly by the buildup of organic materials which release carbon dioxide into the water when they decompose. (Do you recall how carbon dioxide makes water acidic?) The fact that the *p*H of an *oligotrophic* ("little nourishment") body of water generally differs from the *p*H of an *eutrophic* ("adequate nourishment") body is one reason why you may find entirely different populations of plants, animals, and protists in these waters.

If you discover any body of water with a *p*H value beyond the limits 7.0 to 8.5, you should be suspicious of pollution unless you can find a natural cause. Common pollutants that can affect the *p*H of water are the effluents from chemical and fertilizer plants, pulp and paper plants, creameries, and steel mills.

(b) Alkalinity. The alkalinity of a water sample refers to its capacity to neutralize acid. It is caused by bases (alkalis) and basic salts in the water. The most common bases are the hydroxides of sodium, calcium, magnesium, and other metals. The most common basic salts in natural waters are the carbonates and bicarbonates of the same metals. A solution can have a high alkalinity without necessarily being highly alkaline, that is, without having a high *p*H. In other words, a solution may have the potential to neutralize quite a bit of acid without exhibiting a high *p*H. Are you confused? If so, try this experiment.

Place 25.0 ml of 0.1*M* sodium hydroxide solution in a 250 ml Erlenmeyer flask. Determine the *p*H of the solution using Alkacid test paper. Add 0.1*M* hydrochloric acid to the solution, 2 ml at a time. After each addition of acid, mix the solution and determine the *p*H. Continue this process until the Alkacid test paper indicates that the solution is definitely acidic (*p*H = 6.0).

Repeat this entire procedure beginning with 25.0 ml of 0.1*M* sodium bicarbonate solution.

Compare the results of the two experiments. Do you now see that a solution can have a high alkalinity without being highly alkaline?

In North America, alkalinity values are expressed in ppm of $CaCO_3$. This convention is used in the experimental method outlined in Unit 6. The chemical explanation behind the choice of $CaCO_3$ as a standard is complex and is not essential for your pollution studies. Hence, it has been omitted. (The chemical explanation can be found in the section on alkalinity in *Recommended Reading* 5.)

An alkalinity value below 50 ppm is considered low; a value of 200 ppm is becoming quite high. You can expect sewage to have a slightly higher alkalinity than the water in your community. If, however, you discover excessively high alkalinity values in the sewage discharge, you should be suspicious of the dumping of alkaline industrial wastes into the sewage system.

(c) Acidity. As you might expect, a solution can have a high acidity without being highly acidic, that is, without having a low *p*H. Acidity is defined as the ability to neutralize bases. Acidic properties, on the other hand, are due to the hydrogen ions that acids release. Therefore, if a solution contains weak acids (those which do not release all of their hydrogen ions until they are in a reaction with a base), it may have a high acidity because of its *potential* to create hydrogen ions but low acidic properties because it has not yet done so. Typical weak acids are carbonic acid (a solution of carbon dioxide) and organic acids such as acetic acid (vinegar).

Acidity is commonly classified as free acidity and total acidity. The *free acidity* is caused by strong acids like hydrochloric acid and sulfuric acid. This type of acidity can lower the *p*H considerably. Why? The *total acidity* consists of the free acidity plus the acidity contributed by the weak acids present.

(d) Hardness. Hardness in water is caused mainly by calcium and magnesium ions. These ions are generally present in water as sulfates, chlorides, and bicarbonates. In most natural waters the hardness is almost entirely due to bicarbonates, mainly calcium bicarbonate and, to a lesser degree, magnesium bicarbonate. Do you recall how bicarbonates get into natural waters? The presence in water of compounds like calcium chloride and magnesium sulfate depends primarily upon the geology of the countryside around the body of water. If, for example, a stream flows through an area with rock formations consisting largely of gypsum ($CaSO_4$), you can expect that the water will contain hardness in the form of $CaSO_4$ (calcium sulfate). Of course, one cannot rule out pollution as a possible cause of hardness. The calcium chloride ($CaCl_2$) that is frequently used to combat dust on gravel roads and, on occasion, to melt ice on city streets, eventually is washed into streams.

Hardness in water is undesirable for many reasons, most of which are economic. Its economic significance is so great that hardness tests are among the most commonly performed tests of water quality today. The problems are basically these: Calcium and magnesium ions in water react with soap to form curds which hinder effective washing. Further, when these ions are present as bicarbonates, they precipitate out when the water is heated, forming the familiar tea kettle scale and the costly boiler scale that plagues industry. Because of these problems, synthetic detergents were developed that do not form curds in hard water. Water softeners were also installed in many homes and industries.

Hardness caused by bicarbonates is called *carbonate hardness*. It is also called *temporary hardness* since it can be removed by heating. Hardness caused by the sulfates and chlorides of calcium and magnesium is called *permanent hardness*; it cannot be removed by heating. The sum of these two types of hardness is termed the *total hardness* (T.H.) of the water. Hardness values, like those of alkalinity and acidity, are expressed in ppm of $CaCO_3$. The following table shows the commonly accepted standards for degrees of hardness in ppm of $CaCO_3$:

Soft	0–60 ppm
Moderately hard	61–120 ppm
Hard	121–180 ppm
Very hard	over 180 ppm

T.H. values below 250 ppm are considered acceptable in drinking water. Water with a T.H. value above 500 ppm is considered hazardous to human health. In your pollution studies, don't be too quick to blame an industry for high T.H. values. Be sure that you have first studied the geology of the watershed in which the stream or lake is located.

To gain some first-hand experience with water hardness and to reinforce in your mind the ideas in this section, try the following laboratory study.

Place 10 ml of each of the materials listed in Table 4 in a test tube of standard size. To each sample add soap solution drop by drop, shaking after every 2 or 3 drops. Continue to do so until a permanent lather ½ inch in depth forms on top of the solution. Record the number of drops of soap solution required.

Repeat the entire experiment with this change: Heat the material to boiling. Keep it at that temperature for a minute or two. Then cool to room temperature and proceed with the addition of the soap.

TABLE 4

Material	Drops of soap		Other observations
	No heating	Heating	
Distilled water			
Tap water			
$CaCl_2$ solution			
$MgSO_4$ solution			
$Ca(HCO_3)_2$ solution			

Record your results in a table comparable to Table 4.

Are your results consistent with the discussion preceding the experiment? How does this experiment demonstrate carbonate hardness? Permanent hardness? What type of hardness is present in tap water? How does distilled water differ from tap water? Would distillation be a feasible way of restoring water quality?

For Thought and Research

1   Confirm that carbon dioxide imparts acidic properties to water by performing the following exercise. Test the pH of a sample of distilled water using a universal indicator, Fisher Alkacid test paper, or a pH meter. Saturate the water with carbon dioxide and repeat the test.

Recommended Reading 4 describes on page 42 the effect that pH has on the concentration of dissolved oxygen in water. Study this explanation; discuss it, if necessary, with your teacher; then see if you can design an experiment to check on its validity.

2   An industry in Ontario was charged in 1970 with polluting a body of water by discharging acid into it. In court, an official of the company defended its action by stating that the pollution "only amounted to 1 gallon of sulfuric acid for every 10,000 gallons of water." That doesn't sound like very much acid, does it? On a continuing basis, however, it could have a marked effect on the aquatic life. Prepare an explanation of this effect that an ecologically-uninformed company official could understand.

3   High alkalinity is harmful to humans, to most animals, and to many plants. Imagine that an industry discharges a quantity of strongly alkaline waste into a small lake. How might the plant and animal life be affected? Why? If the discharge is stopped, could complete recovery of the aquatic system occur? Explain your answer.

4   Design an experiment to show that a solution can have a high acidity without necessarily being highly acidic. Try to parallel the reasoning and the procedures that were used with alkalinity.

5   (a) How is hardness related to alkalinity, acidity, and pH?
    (b) Can hardness indirectly affect the D.O. concentration of water?

(c) Why is excessively hard water harmful to humans if it is ingested over a long period?

(d) Why do synthetic detergents not form a "bathtub ring"?

(e) How do water softeners work? Does a water softener contribute in any way to water pollution?

Recommended Readings

1 *Chemistry: Experimental Foundations* by R. W. Parry et al., Prentice-Hall, 1970.

2 *Foundations of Chemistry* by E. R. Toon, G. L. Ellis, and J. Brodkin, Holt, Rinehart & Winston, 1968.

3 *Chemistry* by M. J. Sienko and R. A. Plane, McGraw-Hill, 1971.

4 *Concepts of Ecology* by E. J. Kormondy, Prentice-Hall, 1969.

5 *Fundamentals of Limnology* by F. Ruttner, University of Toronto Press, 1963.

2.4 NITROGEN

We have already established the biological importance of carbon, hydrogen, and oxygen. These elements are involved in both photosynthesis and respiration, and are constituents of all living matter. Other elements required in relatively large amounts by most organisms are nitrogen, phosphorus, potassium, sulfur, calcium, and magnesium. Such elements are called *nutrients* by biologists.

One of the most important of the nutrients is nitrogen. It is present in all proteins. Proteins, in turn, are a major component of the planktonic organisms that form the base of all aquatic food webs. Studies have shown that plankton averages about 50% protein; it has a nitrogen content ranging from 7–10%. Because of the importance of nitrogen, we will investigate rather closely its occurrence in nature and the manner by which it is cycled. We will also examine the causes and effects of excessive nitrogen concentrations in aquatic ecosystems.

The Nitrogen Cycle. The occurrence of nitrogen in nature is best studied through an examination of the nitrogen cycle. This biogeochemical, or nutrient, cycle is somewhat complex. Figure 2-4 shows how very complete the cycle is.

Three main reservoirs of nitrogen exist in nature: the atmosphere, inorganic compounds (nitrates, nitrites, and ammonia), and organic compounds (proteins, urea, and uric acid). One would expect the atmosphere, with its huge concentration of nitrogen (78% by volume), to be the most important reservoir. This is not the case, however. Few organisms can make direct

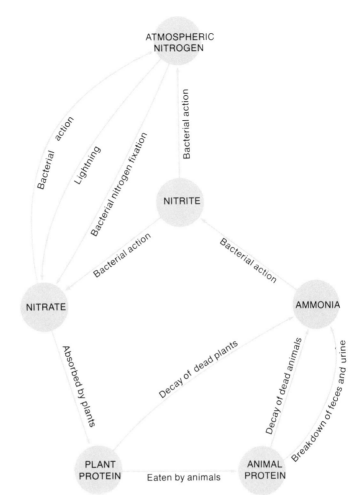

Fig. 2-4
The nitrogen cycle. Follow the nitrogen as it passes from producer to consumer to decomposer and back to producer again.

use of atmospheric nitrogen (N_2). Before most plants can absorb nitrogen, it must be in the form of a nitrate (NO_3^-). This transformation, called nitrogen fixation, may be either chemical or biological in nature.

Lightning causes the most common type of chemical fixation. Bolts of lightning passing through the air provide the energy for oxygen and nitrogen to unite to form nitrogen dioxide (NO_2). This gas reacts with rain water to form nitric acid (HNO_3). It, in turn, reacts with minerals in the soil to form nitrates (NO_3^-). Some of the nitrates in commercial fertilizers are manufactured through processes that are chemically very similar to this.

Most of the nitrate that occurs naturally in soil and in water is formed by biological means. You have probably heard of the nitrogen-fixing bacteria that are located on the roots of legumes. If you have never seen them, carefully remove the roots of a clover, pea, or bean plant from the soil. The nodules on these roots contain bacteria that are capable of fixing atmospheric nitrogen. In a dense stand of clover, they can fix as much as 500 pounds of nitrogen per acre per year! Several free-living forms of bacteria are also capable of converting atmospheric nitrogen into nitrate within the soil. In the aquatic environment, nitrogen fixation is carried out chiefly by a few species of free-living bacteria and by some species of blue-green algae.

Once nitrates have been absorbed by plants, the nitrogen is used to synthesize plant proteins. Then herbivores transform this protein into animal protein. When plants and animals die, decomposer organisms convert the nitrogen in their proteins into the ammonia form (NH_3 and NH_4^+). The nitrogen in fecal matter and urine also ends up as ammonia. The pungent odor of outhouses, chicken pens, and hog yards is ample evidence of this fact. If you are not sure what ammonia smells like, rub some ammonium carbonate between your fingers for a few seconds and then smell it.

Very few organisms can use nitrogen in the ammonia form. Bacteria do exist, though, that convert the ammonia into nitrate. Usually one type of bacterium converts the ammonia to nitrite (NO_2^-); a second completes the conversion to nitrate (NO_3^-). Other bacteria and some fungi convert nitrites and nitrates into atmospheric nitrogen to complete the cycle. It is of interest to note that the nitrogen cycle need not and often does not involve the atmospheric form of nitrogen.

Nitrogen Enrichment—Causes and Effects. As mentioned earlier, there are three main reservoirs of nitrogen: atmospheric nitrogen, inorganic compounds, and organic compounds. Since nitrogen participates in a rather complete biogeochemical cycle, the chances are that all three forms will be present in natural waters. One can expect to find in water:

1) dissolved nitrogen gas;

2) inorganic nitrogen compounds—nitrates, nitrites, and ammonium compounds, as well as ammonia itself;

3) many organic compounds containing nitrogen; for example, the proteins of living and dead organisms and the products of their metabolism such as urea and uric acid.

You should not be alarmed, therefore, if you detect these compounds in water unless, of course, they are present in excess. Which forms of nitrogen can be conveniently used as indicators of nitrogen pollution?

Atmospheric nitrogen can be ruled out. The water is in contact with the atmosphere and, consequently, is normally saturated with nitrogen gas. The organic component is extremely difficult to estimate since one has to measure the nitrogen present in the proteins of living and non-living organisms and in the products of their metabolism. This leaves only the inorganic forms (nitrates, nitrites, and ammonia) as convenient indicators of nitrogen pollution. Let us consider each of these forms in some detail.

(a) **Ammonia** is a by-product of the decay of plant and animal proteins and of fecal matter. It is also formed when the urea and uric acid in urine decompose. Thus the presence of ammonia can be an indication that sewage is entering the water. Since many fertilizers contain ammonia and ammonium compounds, runoff from farms can also contribute to the concentration of ammonia in water.

The nitrogen cycle shows you the role that ammonia plays in natural ecosystems. A further point might interest you. Your municipality undoubtedly chlorinates its water supply. The chlorine kills bacteria that may be present in the water. However, ammonia reacts with chlorine and destroys its germicidal action. Thus when it is present in the water, large quantities of chlorine must be added to reach a concentration sufficient to kill the bacteria.

(b) **Nitrites** are formed in natural waters when ammonia is converted to nitrates by bacteria (see Fig. 2-4). Some nitrites are also formed as nitrates are changed into nitrogen by bacteria. The latter process normally occurs only in stagnant water and, even then, only near the bottom. (Why would this be so?) Since, in both instances, nitrites are intermediate compounds, the concentration of nitrite in natural waters will probably not be appreciable. This is particularly true near the surface. Thus relatively large nitrite concentrations tend to indicate industrial pollution. Nitrites are often used in boiler water to prevent corrosion; if this water is released into the sewage system, nitrite pollution can occur.

(c) **Nitrates** are formed chiefly by electrical storms, nitrogen-fixing organisms, and the action of bacteria on ammonia. All three processes occur independently of man's actions. If, however, man discharges sewage into a body of water, the overall rate of the third process generally increases markedly. This is because sewage liberates ammonia during decomposition. Thus high nitrate concentrations make one suspicious of domestic pollution. Nitrates are present in most fertilizers, so a high nitrate concentration may also mean that farm runoff is a problem.

Other possible sources of increased nitrate concentrations are the natural decay of dead plants and animals, industrial effluents, and animal excreta. With respect to the last source, it has been estimated that a feedlot with 10,000 cattle can cause water pollution problems equivalent to those caused by a city of 45,000 people!

Obviously you will have trouble tracking down the exact sources of nitrates in a stream or lake. One thing is clear, though: nitrates are plant nutrients. If other environmental factors (for example, phosphate concentration) are suitable, nitrates contribute to the aging of a body of water. The ultimate effects on the body of water are the same whether the source of nitrates is natural or induced by man. Only the rates may differ. These effects include an accelerated growth of phytoplankton and other plants, eventually followed by a lowering of the D.O. concentration and changes in the fish population. Foul odors, clogged filters in purification plants, disagreeable tastes, and the lowering of recreational value are other effects.

Acceptable Levels. It is very difficult to give you guidelines for acceptable levels of nitrate and other inorganic forms of nitrogen. What may be an excessively high nitrogen concentration in one body of water may be the normal or expected level in an older body of water that has aged naturally. Further, high concentrations of nitrogen compounds do not always cause eutrophication. The nutrient present in lowest relative amount determines the degree of growth. Thus high nitrogen concentrations may have little effect if, for example, the phosphorus concentration is very low.

Some general guidelines are possible, however. One researcher defines a "well-behaved lake" as one in which total inorganic nitrogen (nitrate, nitrite, and ammonia) does not exceed 0.30 ppm of nitrogen at the time of spring overturn (see *Recommended Reading* 3). *Recommended Reading* 1 points out that concentrations of inorganic nitrogen above 0.30 ppm can be expected to contribute to algal blooms. Both sources indicate, however, that wide deviations from this value occur.

With respect to human health, ammonia in drinking water presents no serious problem since chlorination removes it. Nitrite is very poisonous but not too stable in most waters. Most authorities consider that 45 ppm of nitrate in drinking water render the water hazardous to humans and, indeed, to most animals. The nitrate is changed to nitrite in the stomach. In infants, nitrite in the water can cause methemoglobinemia (a "blue baby" condition). Many deaths occur every year as the result of nitrate poisoning, usually from nitrates in farm well water.

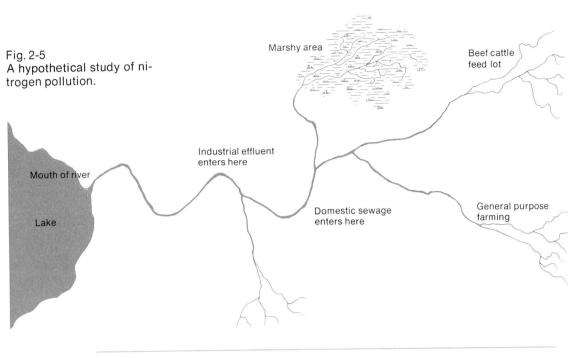

Fig. 2-5
A hypothetical study of nitrogen pollution.

Marshy area

Beef cattle
feed lot

Industrial effluent
enters here

Mouth of river

Domestic sewage
enters here

General purpose
farming

Lake

For Thought and Research

1 Figure 2-5 represents the drainage valley of a hypothetical river. Imagine that you are testing the water quality of this river. At the mouth of the river you obtained these values for the inorganic nitrogen content:

$$\text{nitrate } (NO_3^-) \ = \ 32.0 \text{ ppm}$$
$$\text{nitrite } (NO_2^-) \ = \ 2.5 \text{ ppm}$$
$$\text{ammonia } (NH_3) \ = \ 19.5 \text{ ppm}$$

Select the sites at which you would perform further tests for inorganic nitrogen and state, in relative terms, the anticipated results. The object of these tests is to pinpoint as accurately as possible the source or sources of the pollutants.

2 If you are doing a pollution study on a lake, inorganic nitrogen should be determined when spring overturn is complete but before any significant growth occurs. Why? (Consult *Recommended Reading* 4.)

3 How can the various inorganic forms of nitrogen be removed from water? Consult *Recommended Reading* 1.

4 Use the nitrogen cycle to answer these questions:

(a) What might happen to a lake if the nitrate concentration doubled through pollution over a short period of time?

(b) What would happen to a lake ecosystem if an industry released a relatively large quantity of nitrite-bearing effluent into the water?

(c) Agricultural runoff from fields and manure piles often contains ammonia and ammonium compounds. What effects will this runoff have on lakes that it eventually enters?

Recommended Readings

1 *Cleaning Our Environment. The Chemical Basis for Action*, American Chemical Society, 1969.
2 *Fundamentals of Limnology* by F. Ruttner, University of Toronto Press, 1963.
3 *Readings in Conservation Ecology* by G. W. Cox, Appleton-Century-Crofts, 1969. See articles on eutrophication.
4 "The Nitrogen Cycle" in *Concepts of Ecology* by E. J. Kormondy, Prentice-Hall, 1969.
5 "The Nitrogen Cycle" in *Ecology and Field Biology* by R. L. Smith, Harper & Row, 1966.

2.5 PHOSPHORUS

Phosphorus, like nitrogen, is an element of major significance in the biotic world. Many important organic molecules within cells contain phosphorus atoms. For example, adenosine triphosphate (ATP) is a phosphorus-bearing compound found in every living cell, where it plays a key role in energy storage and supply.

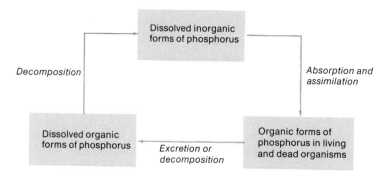

Fig. 2-6
The phosphorus cycle in an aquatic ecosystem.

   The Phosphorus Cycle. Phosphorus, like other nutrients, undergoes a cycle in nature. Its pathway through an ecosystem is so similar to that of nitrogen that we have included here only a summary of the cycle (Fig. 2-6). In an aquatic ecosystem phosphorus is present in three forms:

1)   in inorganic phosphorus compounds, normally orthophosphates (PO_4^{3-}), which are generally referred to as "phosphates";

2)   in organic molecules in the protoplasm of living and dead organisms;

3)   in dissolved organic molecules, most of which are produced by the decomposition of dead organisms and of the waste products of living organisms.

Let us trace briefly the path of phosphorus through a typical aquatic ecosystem. The water normally contains some dissolved phosphates (inorganic phosphorus). Phytoplankton and other plants absorb the phosphate and use it to synthesize molecules like ATP. When herbivores eat the plants, they obtain much of this phosphorus. When the plants and animals die, decomposers return the phosphorus to the water as dissolved organic matter. Additional phosphorus is returned to the water as dissolved organic matter from the excreta of living organisms. Finally bacteria break down the dissolved organic molecules, releasing phosphates that can go through the cycle again. A thorough account of the phosphorus cycle is given on pages 50–54 of *Recommended Reading* 1.

Phosphorus as a Limiting Factor. The ratio of phosphorus to other elements outside an aquatic organism is generally much less than the ratio of phosphorus to the same elements within the living organism. This observation led biologists to conclude that phosphorus is the nutrient most likely to be the limiting factor in eutrophication. In other words, water can have a high nitrate concentration without accelerated eutrophication *provided* the phosphate concentration is very low. Not all biologists, however, agree with this theory. Some believe that the ratio of phosphorus to nitrogen is the limiting factor in eutrophication. One interesting fact has emerged, however. Phosphorus stimulates the nitrogen-fixing activities of the blue-green algae. We can do our best to keep nitrates out of the water, but so long as the water contains phosphates and blue-green algae, our efforts may be of little significance in controlling eutrophication.

Sources and Control of Phosphorus Pollution. Because phosphorus appears to be a limiting factor, much emphasis has been placed recently on the prevention of phosphorus discharge into bodies of water. In a natural or unpolluted body of water, the three forms of phosphorus are in balance. Problems generally arise, though, when outside sources upset this balance.

Phosphorus enters water from many of the same sources as nitrogen: sewage, industrial effluents, agricultural runoff, animal wastes, and decaying plants and animals. As you undoubtedly know, some provinces of Canada and some states of the U.S. began their attack on phosphate pollution by imposing total or partial bans on phosphates in detergents. By the time you read this, a total ban may exist in these two countries. If this is achieved, about 300,000,000 pounds of phosphorus per year will no longer find their way into the rivers and lakes of the U.S. and Canada. Yet over 700,000,000 pounds per year enter our natural waters from other sources, principally in the feces of humans and domesticated animals. That quantity is over 10 times the amount that biologists think is sufficient to promote algal blooms in *all* of the natural waters of North America!

The situation is not hopeless, however, provided our governments act quickly. Tertiary sewage treatment can remove 90–95% of this phosphorus at a cost of about 5 cents per 1,000 gallons of water. Would you like to see how this is done? If so, try this: Prepare a phosphate solution by dissolving a pinch or two of sodium phosphate in 2 inches of water in a test tube. Add calcium hydroxide solution (limewater) or aluminum sulfate (alum) solution until a precipitate forms. What is the precipitate? See if you can remove it from the water. How do you think this precipitate is removed in a tertiary treatment plant? *Recommended Reading* 2, pages 131-132 will help you here.

In Section 2.4 we stressed that runoff from farmers' fields is a major source of nitrogen pollution. Although fertilizers usually contain high concentrations of phosphorus as well as of nitrogen, runoff from fields generally contains very little phosphorus. Apparently many types of soil can "fix" or hold on to phosphate ions. The phosphate ions are adsorbed by the soil particles, that is, they cling to the surface of the particles. Further, the phosphate ions are not released until there is a deficit of them in the solution around the soil particles. If, however, erosion occurs, soil particles together with their films of phosphate ions are carried into streams.

Acceptable Levels. Here again it is difficult to give you precise guidelines, since so many variables enter the picture. One researcher has pointed out that a "well-behaved lake" should not have an inorganic phosphorus content that exceeds 0.015 ppm at the time of spring overturn. (Have you determined the significance of the timing of the nitrate test yet?) Another source says that the annual average concentration should not exceed 0.015 ppm. If the concentration does exceed this value, one can expect algal blooms.

For Thought and Research

1 In 1970 many detergent companies began to replace phosphates in their detergents with a chemical called sodium nitrilotriacetate (N.T.A.). Shortly thereafter this compound was banned from detergents throughout most of North America.

Find out all you can about this compound. What is its formula? Does it break down (degrade) naturally during common sewage treatment processes? Do detergent companies feel that it is as effective as phosphates? Where was N.T.A. first used in detergents? Is it still used there? Does it form nitrates when it decomposes? Has research been conducted to determine its possible effects on aquatic ecosystems? Why was it banned?

2 Find out what kind of sewage treatment facilities exist in your community. Does your community have tertiary treatment that removes phosphates? If not, is it planning to install this in the near future? What are the immediate and long-term plans of your local, provincial or state, and federal governments with respect to phosphate removal through tertiary treatment?

Recommended Readings

1 *Concepts of Ecology* by E. J. Kormondy, Prentice-Hall, 1969.
2 *Cleaning Our Environment. The Chemical Basis for Action*, American Chemical Society, 1969.

2.6 OTHER POLLUTANTS

If you plan to conduct a pollution study of a stream, river, or lake that receives a fair amount of industrial effluent, you should plan to conduct some or all of the following tests: chlorine, chloride, chromium, copper, cyanide, fluoride, hydrogen sulfide, iron, manganese, sulfate, lead, mercury, zinc, and silica.

Before you conduct any of these tests, however, be sure that you know what you are doing and why. Research the topic and be sure you can answer questions such as these:

1)	What is the biological importance of the element or ion?
2)	Does it occur naturally in the aquatic ecosystem that you plan to study?
3)	What effects do excesses of this substance have on aquatic ecosystems?
4)	What is its "normal" concentration in the body of water under study?
5)	What is the maximum concentration that can be allowed without causing harm to the ecosystem?
6)	At what level does the pollutant become hazardous to man?

7)   What are its possible sources? Which of these sources generally contributes the highest proportion of the pollutant?

8)   What measures can be used to prevent the discharge of the pollutant at its source?

9)   How can the pollutant be removed once it enters a body of water?

In Unit 6 you will *not* find suggested methods for determining the concentrations of these pollutants. In general, the standard methods are much too complex. However, Hach and LaMotte kits do exist for most of these tests and we recommend that you use them.

Recommended Readings

For further information, consult:

1   *Cleaning Our Environment. The Chemical Basis for Action*, American Chemical Society, 1969.

2   *Water and Wastewater Analysis Procedures*, Hach Chemical Company, P.O. Box 907, Ames, Iowa 50010. Free publication.

3   *Water Test Kits*, Hach Chemical Company. A free catalog that gives useful background information regarding the pollutants that are under investigation.

4   *Streams, Lakes, Ponds* by R. E. Coker, Harper & Row, 1968. See Chapter 6, "Basic Nutrients in Water."

5   *The Biological Aspects of Water Pollution* by C. G. Wilber, C. C. Thomas, 1969. See Chapter 5, "Metals," and Chapter 10, "Industrial Pollution."

6   *Fundamentals of Limnology* by F. Ruttner, University of Toronto Press, 1963. See "Dissolved Substances and Their Transformations," pages 56-99.

2.7   TOTAL SUSPENDED AND DISSOLVED SOLIDS

You have probably noticed that the appearance of the water in a pond, lake, stream, or river frequently changes. After a heavy rainfall, a river is often brown and silty in appearance, largely from suspended soil particles carried into it by the surface runoff. A lake is often dirty in appearance after a storm due, in part, to wave action stirring up bottom sediments. A river that flows through a city changes in appearance after a rainfall because street runoff and sewage overflow have been added to it (Fig. 2-7). At certain times of year the water of many ponds, lakes, and slow-moving rivers assumes various shades of green and brown, due to algal blooms.

Fig. 2-7
Water from a downtown section of a river before (top) and after (bottom) a heavy rainstorm.

All of these changes in the appearance of the water are due to suspended and dissolved materials in the water. Even under normal conditions such materials are present in the water but, because of their low concentrations, you often cannot see any evidence of their presence.

Freshwater biologists are interested in knowing the total concentrations of suspended and dissolved materials in the water for two basic reasons. First, with the help of such information they can determine the *productivity* of the body of water; that is, its ability to support life. For example, if the dissolved materials are plant nutrients like phosphates and nitrates, the body of water will likely be highly productive with respect to plant life. As a result, it will also be highly productive with respect to animal life. You may recall that such a body of water is said to be *eutrophic*; one that has a low productivity is called *oligotrophic* (see also Section 3.2). Second, by measuring the total concentrations of suspended and dissolved materials at various times of year, at various times of day, and at many locations in a body of water, biologists can establish "norms" for that body of water. Later sudden increases from these norms indicate such things as municipal sewage disposal problems and illegal dumping of industrial wastes. Obviously, then, the measurement of the total concentrations of suspended and dissolved materials is of paramount importance to the person who is studying pollution.

Total Suspended Solids (T.S.S.). The solids suspended in water consist chiefly of living and dead phytoplankton and zooplankton, silt, human sewage, animal excrement, portions of decaying plants and animals, and a vast range of industrial wastes. By definition, the T.S.S. of a water sample is the amount of material, by weight, that is *suspended* (not dissolved) in a given volume of water. Thus T.S.S. values are often expressed in mg per liter. More commonly, though, T.S.S. values are given in parts per million (ppm). For example, a water sample could have a T.S.S. of 92 mg per liter, or 92 ppm, both of which mean that one liter of the water contains 92 mg of suspended solid matter. If you do not see why 1 ppm is equated to 1 mg per liter, look up metric units of mass and volume to clarify the matter. You should then be able to figure out how to determine the T.S.S. of a water sample. Try it. After you have done so, check your method against Section 6.14.

The direct method of measuring T.S.S. (which you undoubtedly came up with) is quite time-consuming. Therefore scientists often measure *turbidity* to estimate the total suspended solids in a water sample. They use an instrument called a turbidimeter that measures the ability of light to pass through the water sample. You have probably noticed how fog (water droplets suspended in air) scatters the light from the headlights of a car, noticeably reducing the intensity of the light. In a like manner, suspended particles in water scatter incident light, thereby reducing the intensity of the transmitted light. A turbidimeter makes use of this fact. This instrument is quite expensive, but if you go on a field trip to a water purification or a sewage treatment plant, you may see one in use.

Acceptable Levels. Since T.S.S. determinations include a variety of solids ranging from clay to living plankton, it is difficult to set "acceptable" standards for T.S.S. However, as pointed out earlier, sudden or even gradual deviations from an established "norm" for a body of water indicate potential problems. A sudden increase in T.S.S. could mean erosion of soil as a result of a heavy rain; it could also mean that the municipal sewage treatment plant was taxed beyond its capacity during a heavy rain. As you are well aware, solids in sewage wastes constitute a much more serious problem than does clay. Being largely organic, they consume great quantities of oxygen during decomposition. How could you determine approximately what fraction of a T.S.S. sample is organic? (Hint: What are the chief products formed when organic compounds like glucose are decomposed through oxidation?)

Total Dissolved Solids (T.D.S.). A T.D.S. determination gives you the total concentration of *dissolved* solids in a water sample. It is normally expressed in mg per liter or in ppm. Thus a T.D.S. value of 200 ppm means that 1 liter of water contains 200 mg of dissolved solids. Among these dissolved solids could be phosphates, nitrates, alkalis, some acids, sulfates, iron, magnesium, most of the pollutants mentioned in Section 2.6, and a host of other substances.

A T.D.S. determination gives a quick assessment of general water quality. The technique is quite simple. You need only evaporate to dryness in a weighed beaker a measured volume of water that has been filtered to remove suspended solids. A final weighing and some simple arithmetic give the T.D.S. value. The weighings should be done to ± 0.1 mg.

Since most dissolved solids impart the ability to conduct electricity to water, a more rapid determination of T.D.S. can be obtained by measuring the *conductivity* of the water. The degree of conductivity is proportional to the T.D.S. in the water. Special instruments called conductivity meters have been developed that read directly in ppm of T.D.S. Watch for them if you go on a field trip to a sewage treatment or water purification plant.

TABLE 5   TOTAL DISSOLVED SOLIDS IN THE GREAT LAKES

Lake	T.D.S.
Superior	60 ppm
Huron	110 ppm
Michigan	150 ppm
Ontario	185 ppm
Erie	180 ppm

Acceptable Values. When the T.D.S. value is below 100 ppm, the body of water is considered oligotrophic. Values above 100 ppm are generally considered to represent eutrophic conditions. Those of you that live along the Great Lakes might find the data in Table 5 of interest.

Biologists consider Lake Erie to be much more eutrophic than Lake Ontario, yet its T.D.S. value is lower. What explanation can you offer for this apparent contradiction?

Lakes with T.D.S. values below 50 ppm are considered "salt-poor." They contain too few nutrients to support a reasonable plant and animal population. A few such lakes exist in Scandinavia. Some Alpine lakes have T.D.S. readings as high as 100–200 ppm and some Baltic lakes 200–400 ppm. The effluent from a municipal sewage treatment plant with secondary treatment facilities contains 300–400 ppm of T.D.S.

These few figures should give you some guidance in the interpretation of T.D.S. data from your pollution studies.

Transparency and Color. Both suspended and dissolved solids affect the transparency and the color of water. Transparency, in turn, is related to the productivity. A low transparency generally signifies a high productivity. Light cannot penetrate very far into the water because of the high concentration of suspended matter. Color in water indicates a water quality too low to permit its use for most industrial and domestic purposes. A detailed discussion of transparency and color is given on pages 12-22 of *Recommended Reading* 3.

For Thought and Research

Your research in this section should revolve largely around *Recommended Reading* 1 and information that you get from your municipality.

1   How effective is each of the following methods of sewage treatment in removing suspended and dissolved solids from water?

(a) primary treatment;

(b) primary plus secondary treatment.

You should first find out what primary and secondary treatment are. Then you will understand the relative effectiveness of these two methods.

2 What kind of sewage treatment facilities does your community have? Inquire as to the average T.D.S. and T.S.S. values of the effluent. Carry out T.D.S. and T.S.S. determinations on the effluent using the methods outlined in Unit 6. How do your values compare with the stated average values? If there is a significant difference between your values and the average values, what might be responsible for the difference?

3 Select one method of advanced waste treatment (for example, tertiary treatment; reverse osmosis; electrodialysis) and research it thoroughly. How does it affect T.S.S. and T.D.S. values? How costly is it? Does your community have it? Is your community considering it? Does it make possible the recycling of waste water for direct reuse? Are any municipalities in your state or province using the method?

4 How are suspended solids removed in water purification plants that prepare water for household use? How are dissolved solids dealt with?

Recommended Readings

1 *Cleaning Our Environment. The Chemical Basis for Action,* American Chemical Society, 1969
2 "Eutrophication of the St. Lawrence Great Lakes" in *Readings in Conservation Ecology* by G. W. Cox, Appleton-Century-Crofts, 1969.
3 *Fundamentals of Limnology* by F. Ruttner, University of Toronto Press, 1963.

2.8   SOME PHYSICAL FACTORS

Water quality can be affected not only by numerous chemical factors, but also by a variety of physical factors. You saw three physical factors in Section 2.7—turbidity, color, and transparency. Other physical factors should also be taken into account during water pollution studies.

Temperature. In Section 2.1 we emphasized the effect of temperature on the dissolved oxygen content of water. Let us look further into this and other effects of excessively high temperatures (*thermal pollution*) on aquatic organisms.

As you know, each species of aquatic organism has its own optimum or preferred water temperature. For example, the optimum temperature for 3 common species of fish are *Cyprinus carpio* (carp), 32°C; *Perca flavescens* (perch), 24°C; and *Salmo trutta* (trout), 15°C. These and other aquatic organisms can tolerate some deviation from the optimum temperature and in many cases can permanently adapt or acclimate to a temperature significantly different from the optimum temperature. However, if the temperature shifts too far from the optimum, the organism either dies or migrates to a new location. With most species of fish, an increase of 5°C in the water temperature can completely disrupt life. This is particularly true if the increase occurs at an unseasonable time of year. For example, if a stream has an average September temperature of 18°C and hot effluent from an industry raises the average temperature to 25°C, a heavy fish kill will occur. If, on the other hand, the temperature of the stream rises to 25°C as a result of normal and gradual summer warming, a fish kill will probably not occur. Why? (Hint: Most species of fish can detect temperature changes as small as 0.05C°.)

Why does an increase in temperature kill fish and other aquatic life? As the water temperature increases, the body temperature of any poikilothermic ("cold-blooded") animal in the water increases. This, in turn, results in an increase in the rate of metabolism within the animal which, of course, increases the oxygen demand of the animal. Yet, as the temperature of the water goes up, its ability to hold oxygen goes down. Eventually a temperature is reached at which the oxygen demand of the animal exceeds the available oxygen, and the organism dies. This temperature is called the *lethal temperature*. Since it is highly unlikely that all members of a particular species will die at exactly the same temperature and time, ecologists usually employ what they call a *tolerance limit median (TLm)* when they are studying the effects of thermal pollution. They normally use either a 24 hr TLm or a 12 hr TLm. If a certain species of fish has a 24 hr TLm of 30°C, then 50% of the fish die within a 24 hour period if the water temperature is 30°C. What does a 12 hr TLm of 35°C indicate?

Elevated temperatures usually increase the toxic effects of chemical pollutants in water. For example, minnows placed in a 0.55 ppm cyanide solution reacted to the cyanide in 72 minutes when the water temperature was 1.2°C; the reaction time decreased to 12 minutes when the temperature was raised to 20°C. This is another example of an important point: It is wrong to consider in isolation the effect of a pollutant on living organisms. Its effect must be considered in conjunction with all of the other environmental factors that affect the organisms as well as each other.

Be sure to make temperature determinations in all of your pollution studies. Be sure, also, to look for and consider both direct and indirect effects of thermal pollution.

Other Physical Factors. Many other physical factors can affect the water quality in a stream, river, or lake. Listed below are the more important ones. You should be able to determine the possible effects of each of these factors on water quality. Think about them carefully and discuss your conclusions with others before you engage in water pollution studies.

For streams and rivers some important factors are velocity of flow, volume of flow, nature of the bottom, and nature of the banks. For ponds and lakes some important factors are ratio of rate of inflow to rate of outflow, depth profile, nature of the bottom, and nature of the shoreline. Procedures outlined in Unit 6 will enable you to study most of these factors in the field.

For Thought and Research

1 Imagine that an industry pours hot effluent into a stream. The effluent raises the average temperature of the stream and the immediate portion of the bay into which the stream flows from 18°C to 30°C. During the period of discharge, the once-common trout vanished from the stream and bay, but carp, seldom seen before, became quite abundant. Would you consider the water to be thermally polluted? Why?

2 Explain how it is possible for a lake or pond to have a rate of inflow that exceeds the rate of outflow with no accompanying change in volume. What change in the T.D.S. value will likely occur with increasing time in such a body of water? Will this body of water be in a state of progressive eutrophication? Why?

3 Figure 2-8 is a photograph of a portion of a river that flows through a large North American city. A sewage plant dumps its effluent into this river a few miles above this scene; several street drains empty into the river above this site. At the time that this photograph was taken, the D.O. immediately above the weir was 0.3

Fig. 2-8
Why was a weir built in this river?

ppm, the T.S.S. was 180 ppm, and the T.D.S. was 340 ppm. A 100 ml sample water contained 20,000 bacteria of fecal origin. What purposes are served by weir? Could this river support a fish population above the weir? Why did the photographer notice the odor of hydrogen sulfide just below the weir? Should this river be posted *"POLLUTED WATER: NO SWIMMING"*? Some of you undoubtedly recognize this scene. Is the river posted?

Recommended Readings

1 *The Biological Aspects of Water Pollution* by C. G. Wilber, C. C. Thomas, 1969. See pp. 166-169 on "Thermal Pollution."
2 *Pollution Probe* by D. A. Chant, New Press, 1970. See pp. 71-73 on "Thermal Pollution."

2.9 SOME BIOCHEMICAL ASPECTS OF WATER POLLUTION

Biochemical Oxygen Demand (B.O.D.). You already know that aerobic decomposer organisms, largely microorganisms like bacteria, are constantly at work in water breaking down organic compounds into carbon dioxide and water. Other bacteria convert ammonia and nitrites into nitrates. These processes, all of which require oxygen, are natural components of biogeochemical cycles and are essential to the functioning of an aquatic ecosystem.

If a body of water contains little organic matter, the aerobic bacteria can break the matter down without upsetting the oxygen balance in the water. Oxygen is replaced by natural means as fast as the aerobic decomposers use it up. If, however, massive quantities of organic matter are present in the water, the population of aerobic decomposer bacteria multiplies due to the additional food. This is usually followed by serious oxygen depletion as the bacteria consume the organic matter. In nature this condition is frequently encountered in marshes, swamps, and at the bottom of stagnant ponds and lakes. In these locations the concentration of organic matter is quite high due, in large part, to the presence of dead plants. The action of aerobic bacteria on this rich source of food often reduces the D.O. level to zero. When this happens, anaerobic decomposer organisms, again largely bacteria, take over as the main decomposers and *putrefaction* sets in. Whereas aerobic bacteria produce odorless water and carbon dioxide, anaerobic bacteria produce gases like methane and hydrogen sulfide—hence the stench. The next time you are near a swamp, marsh, or pond, push a stick into the bottom ooze, pull it out, and smell it. You will be smelling the by-products of anaerobic respiration.

Man has created and is creating anaerobic conditions in countless bodies of water by dumping effluents containing high concentrations of organic matter. Untreated or poorly treated effluent from sewage treatment plants and from pulp and paper mills are two examples. Since the most dangerous consequence of the dumping of organic effluents into natural waters is the depletion of oxygen, scientists have adopted a standard test for measuring the oxygen requirements of an effluent. When the results of this test are compared to the quality of the receiving water, judgments can be made concerning the ability of the latter to handle the organic portion of the effluent without seriously depleting its oxygen supply. This test is called the *Biochemical Oxygen Demand (B.O.D.)* test. It is one of the most commonly performed tests of water quality at sewage treatment plants. The generally accepted B.O.D. test measures the amount of oxygen used up over a 5-day period by aerobic decomposer organisms in a certain volume of effluent at 20°C. The results are commonly expressed in ppm. Thus a B.O.D. of 200 ppm signifies that 200 mg of oxygen are consumed by one liter of the sample over a 5-day period at a temperature of 20°C.

To perform the test, a sample of the effluent is diluted with highly-oxygenated water and a D.O. determination is immediately performed. A portion of the diluted effluent is then placed in the dark at 20°C for 5 days and the D.O. test repeated. The B.O.D. is calculated from the difference between the two determinations. One must, of course, take into account the degree of dilution. Why is the effluent diluted? Why is the sample placed in the dark?

The 5-day time period for the B.O.D. test is a purely arbitrary choice. Few effluents have achieved their total B.O.D. in 5 days because they contain organic compounds that simply do not break down that quickly. Generally speaking, however, 70–80% of the B.O.D. of domestic sewage is satisfied in 5 days. On the other hand, an effluent containing sawdust, wood chips, and bark will register a very low 5-day B.O.D. because of the slow decomposition rate of wood. This does not mean that wood residues are harmless in water; their total B.O.D. is much higher than that of sewage.

Acceptable Levels. It is difficult to determine just what constitutes an acceptable B.O.D. level for the effluent from a sewage plant or industry. It depends on many factors, including the velocity of the receiving stream, the volume per unit time of effluent that is dumped, and the volume of the receiving body of water. Measurements of the B.O.D. of effluents may provide one with an idea of the relative strengths of various effluents but not much more. However, some countries have set up standards of B.O.D. for their waters. For example, the Ontario Water Resources Commission has set 4 ppm as the maximum acceptable B.O.D. of natural waters. This applies to surface water since the B.O.D. at the bottom of even a crystal clear northern lake is normally quite high. In Britain, the standards that have been adopted for river water quality are as illustrated in Table 6.

TABLE 6

General condition of water	B.O.D.
Very clean	1 ppm
Clean	2 ppm
Moderately clean	3 ppm
Of doubtful cleanliness	4 ppm
Poor	5 ppm

Untreated municipal sewage has a B.O.D. of the order of 600 ppm. Primary sewage treatment usually removes up to half of this; secondary treatment removes about 90% of the B.O.D. It may appear that secondary treatment is the answer to the B.O.D. problem. Remember, though, that without tertiary treatment, most of the nutrients like phosphorus and nitrogen still pass into the receiving water. There they stimulate algal blooms and, ultimately, increase the B.O.D. of the water.

At present, one of the best examples of water with a high B.O.D. is Lake Erie. Many cities and small communities are pouring organic effluents into this lake at an incredible rate. Detroit, alone, contributes 500,000 pounds per day of B.O.D. to Lake Erie. (This means 500,000 pounds of oxygen are used up by bacteria as they oxidize the sewage.) The high B.O.D. of the lake has reduced the D.O. level to an average of 0.7 ppm in some locations. At times, the D.O. concentration drops to zero!

Chemical Oxygen Demand (C.O.D.). Although the C.O.D. test is not biochemical in nature, it is included here because of its close relationship to the B.O.D. test.

Many organic compounds do not undergo biological decomposition fast enough to be taken into account by a 5-day B.O.D. test; still others may never undergo biological decomposition. Nevertheless, such compounds can affect water quality in many ways. Thus it is important to have some idea of their concentrations in effluents and in receiving waters. For this purpose, the *Chemical Oxygen Demand (C.O.D.)* test was developed.

You have seen how bacteria oxidize organic matter to carbon dioxide and water. Strong oxidizing agents like potassium dichromate can do the same thing. However, in addition to oxidizing the compounds that bacteria do, potassium dichromate also oxidizes most other organic compounds. As a result, a C.O.D. value is higher than a B.O.D. value of the same water sample.

In the C.O.D. test, a measured volume of the sample is heated with an excess of potassium dichromate solution of known concentration. If sulfuric acid is used as a catalyst, the reaction time is just 2 hours. During this time, the potassium dichromate oxidizes most of the organic matter. This, of course, uses up some of the potassium dichromate. By determining how much is used up, one can calculate the C.O.D. of the sample.

In summary, this test takes into account three things:

1) biodegradable organic matter that would normally be decomposed by bacteria during a 5-day B.O.D. test;

2) biodegradable organic matter that does not decompose in 5 days but, nonetheless, would eventually decompose and affect water quality;

3) organic compounds that are not biodegradable.

Detergents. Those of you who live in or near a big city have probably seen a detergent foam. Although you may find such a scene objectionable from an aesthetic point of view, the foam itself does very little harm to aquatic life. It indicates, however, that the concentration of detergent in the water is probably high enough to harm some forms of aquatic life.

Two general types of synthetic detergent are now common components of the effluent from municipal sewage treatment plants. These are anionic and cationic detergents. Only the anionic are commonly used in products designed for household use. There are two types of anionic detergents. Linear alkyl sulfonates (L.A.S.), often called "soft" detergents, are biodegradable. Bacteria in the environment can break them down. Alkyl benzene sulfonates (A.B.S.), on the other hand, are often called "hard" detergents because they resist bacterial action.

In general, detergents exert harmful effects on fish at very low concentrations. For example, sodium dodecylbenzene sulfonate, a common A.B.S. detergent, progressively destroys the effectiveness of the gills of trout when it is present in water at a concentration of just 5 ppm. (Raw municipal sewage contains, on the average, 10 ppm of detergent.) At concentrations as low as 0.5 ppm, many hard and soft detergents destroy the taste buds of certain fish species in 3 to 4 weeks. At a concentration of 10 ppm, this destruction is accomplished in only 1 day. You can imagine the effect that lack of taste buds will have on the feeding habits of fish! How often do streams or rivers carry a concentration of detergent high enough to do this? As an example, the portion of the Illinois River downstream from Chicago generally has a concentration of over 0.5 ppm of detergent for more than 150 miles.

Here are a few more facts and figures that might be of interest to you. The 30-day TLm of A.B.S. detergents for one species of fish, the bluegill, is 15–18 ppm. This means that 50% of bluegills die within 30 days when the concentration of A.B.S. is around 15–18 ppm. At concentrations of 13 ppm, the rate of growth of these fish is inhibited and, of course, gill damage is extensive. It has also been discovered that detergents decrease the tolerance of fish to low D.O. levels. They further compound this problem by making it more difficult for oxygen to enter water at the surface.

Aquatic plants also suffer if detergent concentrations are too high. Kelp, for example, cease all photosynthetic activity when the concentration of one A.B.S. detergent is 2 ppm for 2 days.

Acceptable Levels. One research study indicated that most fish species show no serious adverse effects over a 30-day period when the detergent concentration is as high as 3 ppm. One must consider, though, the effects that such a concentration has on the organisms that the fish eat.

A maximum allowable concentration for drinking water is usually considered to be 1 ppm. Since the North American detergent industry is switching from hard to soft detergents, one would expect little or no detergent to find its way into the domestic water supply. The soft detergents should break down before finding their way to water used for drinking purposes. There are a few problems here, though. The soil around many sources of drinking water, for example, at summer cottages, is saturated with hard detergents from the septic system. Detergents are quite mobile in soil so they can eventually reach the ground water used for drinking. Further, it is possible even for soft detergents to reach the drinking water supply without being decomposed. Very little decomposition occurs below the first inch or two of soil.

Why? And where are the tiles of a septic system located? Thus detergents that have not been decomposed in the septic tank can seep through the soil to return in drinking water.

The Solution. The first step is to replace hard detergents by soft detergents. This conversion is well under way in North America now. Remember, however, that even though the health hazard is reduced by soft or biodegradable detergents, eutrophication is enhanced. Many plant nutrients are released when the detergents decompose. Proper advanced treatment methods can remove 90% of these nutrients and any undecomposed detergents from sewage water.

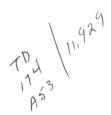

Many housewives have switched from synthetic detergents to soaps for laundrying purposes in order to lessen environmental problems. Soaps, too, are toxic to fish. The 96-hour TLm for minnows is 34 ppm of soap in soft water. In hard water this rises to over 1,000 ppm. Soaps do, however, undergo bacterial decomposition very rapidly. Further, the by-products of this decomposition are not as nutritious for plants as are the by-products of detergent decomposition.

For Thought and Research

1 Interpret the following data:

(a) A sample of untreated municipal sewage has a B.O.D. of 600 ppm and a C.O.D. of 850 ppm.

(b) A river has a B.O.D. of 6.5 ppm immediately below an inlet from a sewage treatment plant. Three miles downstream the B.O.D. is 1.2 ppm.

(c) A sample of untreated municipal sewage has a detergent concentration of 11.2 ppm. A few days later, the sample was re-tested and found to contain only 1.5 ppm of detergent.

2 Refer to Figure 2-9 as you read this description:

Graph A is a plot of the volume of flow of a river against time in months. Graph B is a plot of B.O.D. versus time in months. The data were obtained at a site *upstream* from a town that dumps its sewage residues into the river. Graph C is also a plot of B.O.D. versus time. Here, though, the data were obtained at a site *downstream* from the town.

Study the graphs carefully and account for their shapes.

3 Does your community monitor on a regular basis the B.O.D. of its untreated sewage and the B.O.D. of the effluent from the treatment plant? If so, what are the values? Is the effect of the effluent on the B.O.D. of the receiving waters regularly monitored? What is the effect? If your community does not monitor B.O.D. levels, why doesn't it?

4 Suffolk County in Long Island, New York, has a population of 1,200,000. It has no major sewage system and, therefore, depends almost entirely on septic tanks and cesspools to handle domestic sewage. In the late sixties, drinking water came from taps with a foul odor, a disagreeable taste, and an unsightly foam. On March

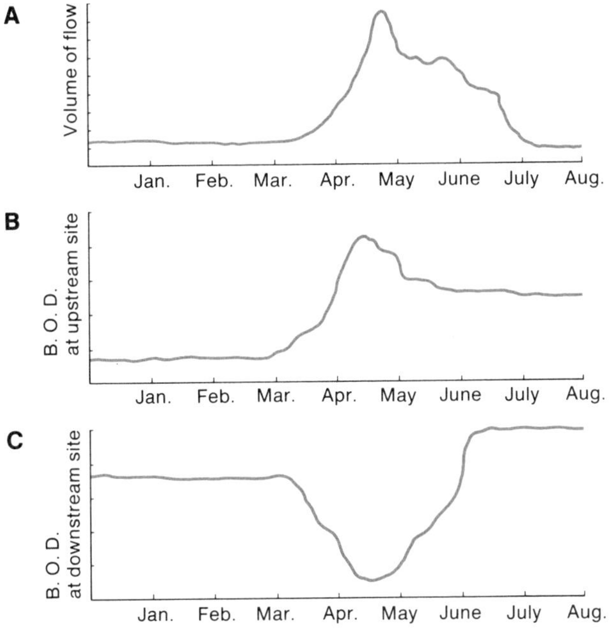

A Volume of flow

Jan. Feb. Mar. Apr. May June July Aug.

B B. O. D. at upstream site

Jan. Feb. Mar. Apr. May June July Aug.

C B. O. D. at downstream site

Jan. Feb. Mar. Apr. May June July Aug.

Fig. 2-9
Account for the shape of each graph and for the shapes of the graphs relative to one another.

1, 1971, the sale of virtually all detergents for home use was banned in Suffolk County. Explain why this legislation was necessary. Should improved water quality have been noticed immediately following the ban? Why? The ban applies to the sale of detergents but not to the use of detergents. What do you think might happen as a result of this?

Recommended Readings

1 *Cleaning Our Environment. The Chemical Basis for Action*, American Chemical Society, 1969. See the sections on B.O.D., C.O.D., and detergents.
2 *The Biological Aspects of Water Pollution* by C. G. Wilber, C. C. Thomas, 1969. Pages 145-151 discuss thoroughly the effects of detergents on fish.
3 *Pollution Probe* by D. A. Chant, New Press, 1970. See the section on "Organic Pollution," pp. 49-54, for an excellent discussion of B.O.D.

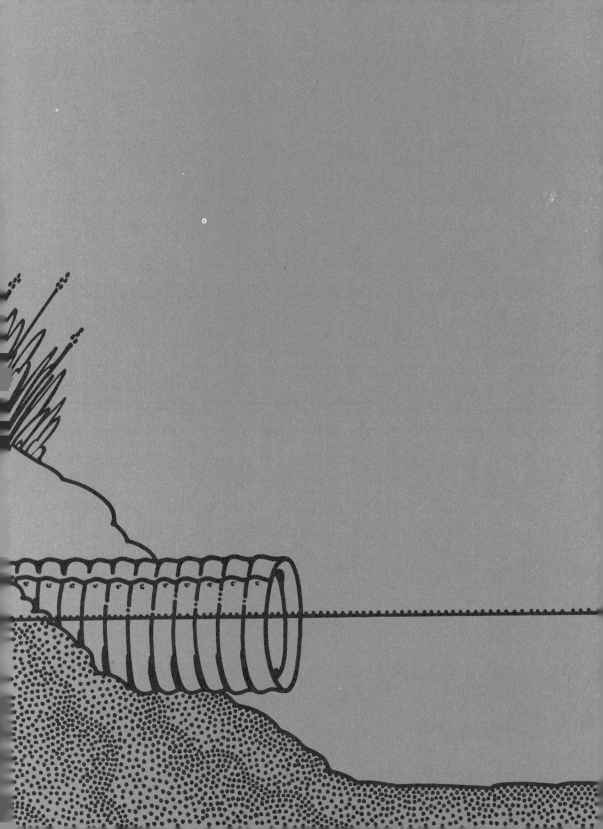

Biological Aspects of Water Pollution

3

By now you are well aware that the addition of chemicals, sewage, and other pollutants to natural waters will, in some way, affect the living organisms that inhabit those waters. Just how the organisms are affected is the main topic of this unit.

As we examine the biological aspects of water pollution, keep in mind that you must not be too hasty in drawing conclusions based solely upon biological data. It is conceivable, for example, that some of the fauna which thrive in a polluted lake might also be the normal fauna of a completely unpolluted lake only a mile or so away. Therefore, using biological material as the only measure of pollution can sometimes be very misleading. Chemical tests like those discussed in Unit 2 are required along with the biological data to give an overall picture of the water quality.

3.1 INDEX SPECIES

A pollutant can be present in water in a concentration high enough to be toxic to all living organisms in that ecosystem. More commonly, though, the pollutant is present in a concentration that kills certain species without harming others. It is also possible for a pollutant to actually promote the growth of certain species. (Can you recall some examples from Unit 2?) Thus when water becomes polluted, there tends to be a shift from

62

many species of moderate population to a few species of high population. Such a decrease in the diversity of the species present is often considered to be an indication of pollution. The species present in the highest concentrations are called the *index species* or the *indicator organisms* of pollution.

Lake Erie once contained such fish as trout, char, chub, whitefish, and lake herring. Over the past 50 years these species have been greatly reduced in numbers. The fish population now consists mainly of carp, perch, and smelt. Apparently these fish can tolerate the low oxygen conditions brought about by pollution of the lake. They can be considered, therefore, to be index species of pollution in that body of water. Lack of any fish at all in Lake Erie would indicate an even higher degree of pollution.

Fish are difficult to use as pollution indicators, however, unless one is doing a prolonged study. Other aquatic species which are less mobile are more often used as biological indicators. For example, algae can be used. Lake Waukasa, Wisconsin, contained prior to 1951 at least 10 different species of algae in its algal blooms. From 1951 to 1958 Lake Waukasa received secondary effluent from the Madison sewage treatment plant. During most of this period 99% of the algal blooms consisted of an alga of the genus *Microcystis*. Since 1958 the effluent has been diverted around the lake. Shortly after the diversion *Microcystis* ceased to be the predominant alga present but became only one of many. In this example *Microcystis* is classified as an index species of pollution.

In addition to the particular fish and alga just discussed, there are many other flora (plants) and fauna (animals) that serve as biological indicators of pollution. After Section 3.2, these index species will be discussed under the broad categories of bottom fauna, bacteria, algae, zooplankton, and fish. You need not remember the species names, but you may require them during your studies. For the moment, just remember that the index species present are dependent upon two things—the nature of the pollutant and the stage of eutrophication of the particular body of water. Since two factors are involved, it is virtually impossible to prepare a list of index species that tells us with certainty the nature and the degree of pollution. Therefore biological indicators must not be used alone.

3.2 EUTROPHICATION

Lakes are essentially transitory bodies of water which eventually fill up and disappear. On a geological time scale they do this very quickly. Yet, during your lifetime, you are not likely to see any marked changes in a lake that is untouched by man and his by-products. Most of the lakes in eastern North America resulted from the last ice age. When the glaciers retreated they left behind basins which filled with water and became lakes. The largest of the glacial lakes are the Great Lakes. Lakes formed in this manner are cold, clear bodies of virtually sterile water. They contain little or no life and are called *oligotrophic* (little nourishment) lakes. Gradually the streams which feed these lakes import nutrients like phosphates and nitrates. These promote the growth of some aquatic organisms whose numbers increase markedly. In addition, silt and dead organisms are carried by the streams into the lakes and are deposited on the bottom. Lake organisms that die are added to the bottom sediment. As the living matter increases and as the inorganic and organic wastes accumulate on the bottom, the lakes become smaller and shallower. The sides of the lakes may also erode to some extent, thereby helping to fill in the basin. As the depth decreases, the water becomes warmer and plants begin to root in the bottom. The combination of higher temperatures and decreased depth greatly increases the amount of life in the lake. Ecologists say that the *productivity* of the lake has increased. At this stage the lake is called a *eutrophic* (adequate nourishment) lake. The process of aging or increasing productivity is referred to as *eutrophication*. In general, a lake is considered to be eutrophic when its average T.D.S. value exceeds 100 ppm.

The plant and animal life changes markedly as eutrophication progresses. For example, as the lake becomes shallower and warmer, the fish shift from those preferring cold water to those which like shallow, warmer water. Thus trout, whitefish, and walleye give way to perch, bass, and sunfish (Fig. 3-1). Some recent studies conducted in Lake Erie illustrate these changes

Fig. 3-1
Succession of fish populations. Inhabitants of oligotrophic lakes include trout (A), whitefish (B), and walleye (C). As lakes age, these fish are replaced by fish such as perch (D), largemouth bass (E), and sunfish (F).

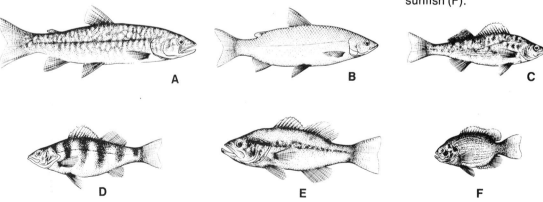

A   B   C

D   E   F

Fig. 3-2
Changes in fish popula-
tions in Lake Erie.

Fig. 3-3
D.O. versus depth for an
oligotrophic (A) and a eutrophic (B) lake at the time
of summer stagnation.

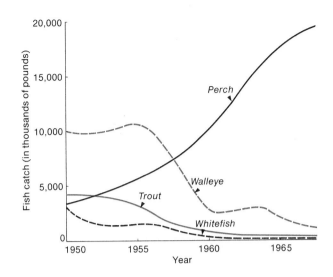

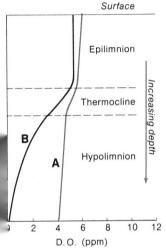

(Fig. 3-2). Such rapid changes cannot be explained by natural eutrophication alone, since it is a very slow process in a lake as large as Lake Erie. Indeed, the amount of natural eutrophication in a typical lake is almost imperceptible in a human lifetime. Yet we see drastic changes in fish populations in Lake Erie over the past 20 years. Obviously the natural process of eutrophication is being accelerated considerably by pollution. The effects of pollution in the water body depend upon the volume of the water present. Of the Great Lakes, Lake Erie is the smallest and has the greatest human population density around it. Thus the effects of pollution are the most evident in it.

For Thought and Research

1 A small lake is found to contain mainly carp, catfish, and other fish that can tolerate low oxygen conditions. Is the lake polluted? Explain your answer.
2 Figure 3-3 shows the variation of dissolved oxygen with depth during the period of summer stagnation in an oligotrophic lake and a eutrophic lake. Account for the shapes of the two curves.
3 "Water pollution is fundamentally a biological phenomenon." Do you agree or disagree with this statement? Why?
4 "The combined efforts of teams of biologists and chemists constitute our best hope for the solution of the problems of pollution." Do you agree or disagree with this statement? Why?

Recommended Readings

1 *Concepts of Ecology* by E. J. Kormondy, Prentice-Hall, 1969. See pp. 180-191, "Water Pollution and the Great Lakes."

2 "The Aging Great Lakes," *Scientific American*, November 1966.

3 *A Guide to the Study of Freshwater Ecology* by W. A. Andrews et al., Prentice-Hall, 1972. See pp. 26-28.

4 *The Biological Aspects of Water Pollution* by C. G. Wilber, C. C. Thomas, 1969. Chapter 12, "Indicator Organisms," gives a detailed account of index species.

5 *The Biology of Polluted Water* by H. B. N. Hynes, Liverpool University Press, 1960.

6 *Readings in Conservation Ecology* by G. W. Cox, Appleton-Century-Crofts, 1969. Articles 30 and 31 discuss index species and eutrophication.

3.3 BOTTOM FAUNA

The types of fauna found on the bottom of a stream or river depend to a large degree on the nature of the bed. This, in turn, is directly related to the current speed of the water. In rapidly flowing areas of a stream or river, all but the larger rocks are gradually transported downstream. As the current slows down, stones and then gravel are deposited. Where the current is still slower, the bottom is sand. Finally, in pools and other places where there is little or no current, silt and mud are deposited on the bed. A more quantitative picture of the relationship between current speed and nature of the bed in a typical stream or river is given in Table 7.

TABLE 7

Current speed	Nature of bed
Over 1 meter/sec	rock
60–100 cm/sec	stones and heavy gravel
30–60 cm/sec	light gravel
20–30 cm/sec	sand
10–20 cm/sec	silt
Under 10 cm/sec	mud

You know, of course, that the current speed is not uniform over the entire length of a river. What factors cause it to vary?

You saw in Unit 2 how two other factors, *temperature* and *dissolved oxygen*, play important roles in determining the fauna present in a body of water. Since the water in a fast-moving stream is somewhat turbulent, it will, as a rule, be highly oxygenated. Thus the bottom fauna are those which require high oxygen conditions. They must also have physical features that enable them to cling to the bottom; otherwise they would be swept downstream. Some of the more predominant bottom fauna of this area of a stream are the nymph stages of such insects as stoneflies and some species of mayflies, as well as the larval stages of some species of caddisflies. (See Sections 3.2 and 4.3 of *Recommended Reading* 1.) All three are generally found under and around stones. The dragonfly nymph also requires clean oxygenated water, but it is normally found among clusters of aquatic plants where the current is not so strong.

Some species that inhabit mud-silt bottoms where the oxygen concentration is low are shown in Figure 3-4. The sludge worm, *Tubifex*, is a segmented worm of the family Tubificidae. It operates on the muddy bottom in much the same manner that its terrestrial cousin, the earthworm, operates on soil. You can easily identify *Tubifex* by its red color and its continuous spiral motion. It constructs a tube above a burrow extending down into the bottom materials. While it feeds, mouth down into the bottom, its tail sticks out of the tube swaying back and forth, exchanging carbon dioxide for any available oxygen in the water. Many species of *Tubifex* exist and they seem to inhabit polluted waters in definite patterns. Some researchers now believe that these patterns are determined largely by the types of bacteria in the bottom ooze. *Tubifex* feed on these bacteria and, apparently, different *Tubifex* species have different tastes.

The bloodworm is another common bottom burrower that can tolerate low oxygen conditions. This blood red organism is actually not a worm, however. It is the larval form of the midge fly, *Chironomus*. These larvae normally occupy tubes that they manufacture from sludge material. Frequently the tubes are found in heaps. The adult midge is an annoying little creature that resembles a miniature mosquito without piercing mouthparts.

Leeches and aquatic sow bugs (isopods) generally frequent muddy areas with slightly higher oxygen concentrations. This is also true of some species of mayfly nymphs and caddisfly larvae. You might note that the fauna which inhabit mud-silt bottoms of streams are also among those that you will find on the bottom of many lakes and ponds. Why?

In polluted waters, changes occur in the chemical and physical characteristics of the water which, in turn, cause changes in the bottom fauna. If the pollution is light, we may simply find small changes in population sizes of the species present. If the pollution is a little more severe, some species of some families may vanish and a few new species may appear. In heav-

Fig. 3-4
Some typical bottom fauna of waters having low oxygen concentrations.

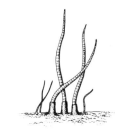

Tubifex, the sludge worm

Chironomus, the midge larva

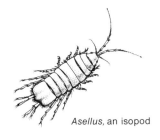

Asellus, an isopod

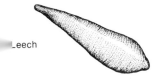

Leech

ily polluted waters, gross changes in the fauna occur. Whole families rather than individual species are usually affected. In general, the greater the pollution, the greater is the change in the types of organisms. In this sense, then, bottom fauna are very useful biological indicators of pollution. Let us take a more detailed look at the effects of pollutants on bottom fauna.

(a) Dissolved Solids. Many dissolved solids are, of course, toxic to bottom fauna if they are present in high enough concentrations. Whether or not a particular substance is toxic depends not only upon its nature, but also upon the ability of the animals to metabolize and eliminate it. Different animals can tolerate different concentrations of a particular substance due to their varying abilities to metabolize or otherwise deal with the substance. The toxicity of a substance also depends on many other factors such as temperature, dissolved oxygen, and pH.

The concentration of any substance will usually be the greatest at the effluent site. As the effluent moves downstream it is diluted and the concentration of any toxic substance decreases. Thus the effect on the bottom fauna should decrease also. Watch for trends of this type in your pollution studies. Be sure, though, that you perform the necessary chemical tests to determine what substances are responsible for the trends that you observe in the bottom fauna.

(b) Inorganic Suspended Solids. Chalk, gypsum, and powdered mine wastes are typical inorganic solids often found suspended in polluted waters. Since they tend to make the water opaque to light, they retard photosynthesis and, as a result, reduce the total biomass of producers. Since the producers are either a direct or indirect source of food for the bottom fauna, the fauna also tend to decrease. Further, when the suspended solids settle, they frequently smother most of the fauna. Near an effluent of inorganic solids one is likely to find only burrowing forms like *Tubifex* and the *Chironomus* larva. Downstream from the effluent site the amount of suspended solids decreases due to gradual settling. Gradually the normal fauna reappear as shown in Figure 3-5.

(c) Heated effluents are becoming more of a problem as the number of power generating plants increases. Power plants use water, generally from a nearby lake, as a cooling agent. The water is returned to the lake at a temperature significantly higher than when it entered the plant. A rise in temperature of only a few degrees can alter the composition of the animal communities that live in the location. Bottom fauna respond to elevated temperatures much as fish respond (see Section 2.8). Their metabolic rates increase markedly, often to the point where they literally

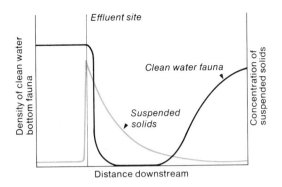

Fig. 3-5
Relationship between
suspended solids and
clean water bottom fauna
above and below an
effluent site.

burn themselves out; or, they die because they cannot obtain enough food and oxygen to sustain the higher metabolic rate.

(d) Organic effluents constitute a serious form of pollution since they have a marked effect on the oxygen content of the water. They include suspended organic matter from domestic sewage, wood chips from pulp and paper operations, and many dissolved organic compounds. As you know, bacteria decompose many of the compounds in organic effluents. This decomposition requires oxygen. Therefore, just below the effluent site there is a sharp decrease in the D.O. concentration of the water. This decrease is directly related to the concentration of organic material in the effluent. As the effluent moves downstream it is gradually diluted and its B.O.D. is reduced. In addition, aeration and photosynthesis add oxygen to the water. If the original input of effluent is not too high, complete recovery of a stream can occur (Fig. 3-6). The rate of re-oxygenation can be affected by the type of organic effluent as well as by its concentration. For example, oils and detergents can slow down and even inhibit oxygen exchange between the air and water.

When the concentration of organic matter is high enough to totally de-oxygenate the water, the normal water fauna are sometimes completely displaced. Further, if the current is swift,

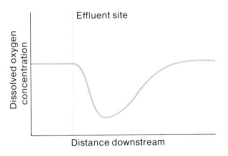

Fig. 3-6
The effect of organic
effluents on the dissolved
oxygen concentration.

few other organisms can live in this region. However, if the current is very slow or if the effluent is discharged into a lake, new organisms move into the region of de-oxygenated water. The larvae of *Culex* (the mosquito), *Psychoda* (the moth fly or sewage fly), and *Eristalis* (the robber fly, hover fly, or bee fly) are the predominant new species (Fig. 3-7). Strangely enough, these larvae have high oxygen requirements even though they can inhabit water with zero D.O. They obtain their oxygen directly from the air through air tubes protruding from their bodies. In unpolluted waters, they usually occupy the sides of streams, lakes, and ponds where organic material collects. Their population is normally kept down by other fauna which eat them. In polluted water or in extremely organic water like marshes, the predators are unable to survive the low oxygen conditions. Without the predators to keep the population down, *Psychoda*, *Culex*, and *Eristalis* multiply rapidly. Thus their presence in large numbers indicates highly de-oxygenated water. *Psychoda* can even tolerate the conditions present in household drains; both *Psychoda* and *Eristalis* thrive in stagnant, foul sewage water. Remember, though, that the cause of de-oxygenation can be natural. These species are not, therefore, absolute indicators of pollution.

In regions where the organic content does not completely de-oxygenate the water, other organisms take over as the predominant species. For example, when the D.O. level is around 15% of saturation, sludge worms (*Tubifex*) are usually quite common. (See Figure 3-4.) Since the mouths of rivers generally have low oxygen conditions due to their highly organic silt bottoms, sludge worms are usually present there. If the water is heavily polluted with organic effluent, these worms occur in abnormally high numbers downstream from the effluent. Recent studies at the mouth of the Don River in Toronto, Ontario, have revealed populations of sludge worms in excess of 500,000 per square meter!

Where the oxygen concentration is a little higher, the sludge worms are usually accompanied by bloodworms, the larvae of the midge fly, *Chironomus* (Fig. 3-4). Since bloodworms can tolerate mild oxygen deficits, they are normal inhabitants of non-polluted silt- and mud-bottom streams and lakes. They thrive in water polluted with organic matter because, like sludge worms, they eat organic matter. Since they are not as tolerant of low oxygen conditions as are sludge worms, they are frequently found in regions of streams where recovery from organic pollution is underway.

Summary. The organisms described here are by no means all that are present in polluted waters. They are only some

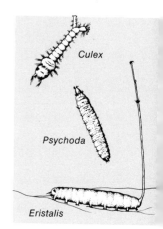

Fig. 3-7
Larvae which may occur in calm regions of water that is heavily polluted with organic matter.

Culex

Psychoda

Eristalis

of the major index bottom fauna that are commonly present. No single one of these index species, by itself, gives an accurate measure of pollution.

For Thought and Research

1 Prepare a list of bottom fauna in order of increasing ability to tolerate low oxygen conditions. Retain the list for use during field work.

2 Three decades ago mayflies were so abundant in some towns and cities bordering Lake Erie that shovels were used to remove their bodies from streets and sidewalks. This problem no longer exists. Why?

3 Figure 3-8 shows the effects of sewage effluent on some stream organisms. Interpret the graph, paying particular attention to the relationship between distance downstream and the relative numbers of each organism.

4 In Figure 3-9 each species of bottom fauna is represented by a number. Its relative abundance at each study site is indicated by the number of times its number appears in the box. Interpret the data.

Fig. 3-8
Effect of organic pollution on some bottom fauna.

Fig. 3-9
The effect of organic pollution on the kinds and abundance of bottom fauna. The numerals in the boxes represent the species present. The abundance of each species is represented by the number of times its numeral is repeated in the box.

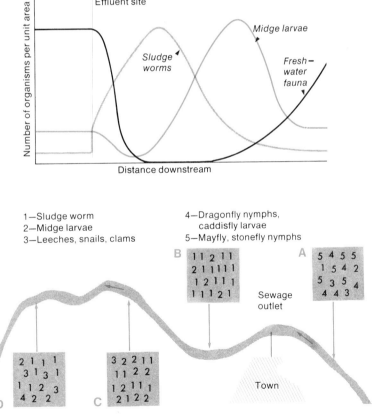

1—Sludge worm
2—Midge larvae
3—Leeches, snails, clams
4—Dragonfly nymphs, caddisfly larvae
5—Mayfly, stonefly nymphs

Recommended Readings

1 *A Guide to the Study of Freshwater Ecology* by W. A. Andrews et al., Prentice-Hall, 1972.
2 *Concepts of Ecology* by E. J. Kormondy, Prentice-Hall, 1969. Read "Water Pollution and the Great Lakes," pp. 180-188.

Consult the following books for the identification of organisms that are not discussed in this book.

3 *The New Field Book of Freshwater Life* by E. B. Klots, G. P. Putnam's Sons, 1966.
4 *A Guide to the Study of Fresh-Water Biology* by J. G. Needham and P. R. Needham, Holden-Day, 1962.
5 *Pond Life* by G. K. Reid, Golden Press, 1967.

3.4   BACTERIA

Bacteria are classified into four major groups according to their feeding habits. *Saprophytic* bacteria feed on dead organic matter. They cannot produce their own food and as a result must rely entirely upon organic compounds present in the environment. As you know, these bacteria occupy the niche of decomposer in an ecosystem. They obtain energy by breaking down organic matter into simple compounds. Without them the surface of the earth would soon be covered to an incredible depth with the remains of dead plants and animals.

Commensal bacteria inhabit living systems but are not parasitic upon them. They take food from the system but also serve a valuable function in that system. For example, bacteria present in the large intestine of humans assist in the breakdown of organic material into simpler constituents. They do no harm to humans. In fact, without them we would all soon die. In return, the intestine provides the bacteria with a good environment. The temperature is controlled, and moisture and food are abundant. So suitable is the environment that these bacteria cannot exist for long outside of the intestine.

Parasitic bacteria also get their food from living hosts. In this case, however, the host usually suffers. Among the parasitic bacteria are the *pathogens* or disease-producing bacteria. Some of these bacteria secrete substances called toxins; the toxins cause the disease symptoms of such diseases as scarlet fever, lockjaw, whooping cough, typhoid fever, and diphtheria. In these diseases, the toxin is spread through the host's body by the circulatory system. As a result, the symptoms of the disease often

Fig. 3-10
Characteristic shapes and
arrangements of bacteria.

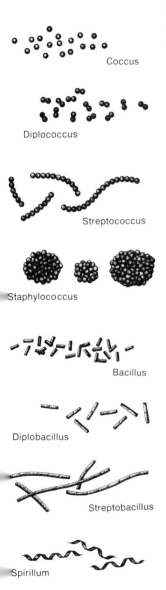

Coccus

Diplococcus

Streptococcus

Staphylococcus

Bacillus

Diplobacillus

Streptobacillus

Spirillum

Vibrio

appear far from the point of entry of the bacteria. Other parasitic bacteria do, however, attack and digest specific tissues. For example, meningitis bacteria live only on the lining of the brain and spinal cord, and tuberculosis bacteria devour lung tissue.

The last type are the *autotrophic* bacteria. They can synthesize their own food from simpler inorganic substances. A few of these bacteria are photosynthetic. This means, of course, that they have a chlorophyll-like pigment.

Shape and Size. Although there are thousands of different species of bacteria, the individual organisms commonly have one of three general forms: ellipsoidal or spherical, cylindrical or rod-shaped, and spiral (Fig. 3-10).

Spherical and ellipsoidal cells are designated as *cocci* (singular *coccus*). These are arranged in different configurations which can be used for identification of the bacteria. If they appear in pairs they are called *diplococci*; when they grow in chains they are called *streptococci*; when they grow in large grape-like clusters they are called *staphylococci*.

Cylindrical or rod-like bacterial cells are designated as *bacilli* (singular *bacillus*). They do not have the variety of groupings that the cocci have, but they do occasionally appear in pairs (*diplobacillus*) and in chains (*streptobacillus*).

Spiral-shaped bacteria are called *spirilla* (singular *spirillum*). These are usually unattached individual cells. Short incomplete spirals, known as *vibrio* or *comma* bacteria, also exist.

Remember that these are only the major configurations. Shapes can be helpful in identifying bacteria, but since there are so many bacterial species and so few shapes, other methods are required. Bacteria are now identified mainly with biochemical procedures. Examples are the ability of the bacterium to grow on certain media and to produce certain gases.

The unit of bacterial measurement is the micron (μ) which is equivalent to 0.001 mm. The majority of bacteria measure approximately 0.5–1.0 μ. Bacilli have a length of 2–3 μ. Some filamentous bacteria are as long as 100 μ.

Bacteria and Pollution. Saprophytic bacteria play a very significant role in water polluted with organic material. As these decomposer organisms break down organic compounds, they use up great amounts of oxygen. The lower oxygen concentration, in turn, can seriously affect the fauna present.

Autotrophic bacteria infest inorganic effluents. Mine drainage contains iron and sulfates. Sulfur- and iron-oxidizing bacteria attack these compounds producing end-products that lower the *p*H of the water. This high acidity destroys most fauna and flora.

Of greatest concern to environmental scientists are those bacteria that inhabit human sewage. Human excreta in the effluents from sewage treatment plants are the greatest source of pathogenic bacteria in our waters. Pathogenic bacteria can also enter the water by leaching through the soil from cesspools and privies, especially in cottage areas where the soil is shallow. One might assume that these bacteria can be detected easily by analyzing water samples. This, however, is not the case. Survival of pathogenic bacteria after leaving the human body is poor, since they require body temperature and fluids to survive. Some of them do, however, remain alive for a limited period of time, although they cannot reproduce. These few survivors are very difficult to detect. The best methods are tedious and expensive. Further, pathogenic bacteria enter the water only sporadically. This increases the chances that they may escape detection. For these reasons, another method is used to indicate the presence of pathogenic bacteria.

Certain species of a group of bacteria called *coliforms* are normal inhabitants of the large intestine of man and other animals. These species are referred to as *fecal coliforms*. Non-fecal species are frequently found in soil and in plants. The coliforms are grouped together because of their similar biochemical reactions. The fecal coliforms have a commensal relationship with the host animal. They are provided with a suitable environment in the large intestine. They, in turn, assist the host in the breakdown of waste products of digestion. Each day billions of coliform bacteria leave the body of each human in the feces. Of these coliforms, 80-95% are commonly of the species *Escherichia coli* (*E. coli*). Thus the presence of coliform organisms, in particular *E. coli*, indicates pollution of the water by fecal matter. Since the feces of a diseased person could contain pathogens, the presence of *E. coli* indicates the *possible* presence of pathogens. Pathogens are hard to detect; coliforms are not. Therefore public health officials look for coliforms when they are assessing water quality. Coliforms live much longer outside the human body than do most pathogens. This fact increases their usefulness as indicators of infected water. If coliforms are not detectable in the water, it is likely free of pathogens and, therefore, fit to drink.

E. coli are usually found as rods about 0.5 μ in diameter and 1.0–3.0 μ in length. They also appear occasionally in a somewhat coccoid form; they may occur singly, in pairs, and in short chains. Obviously direct observation is of little value in the detection of these organisms. They are usually identified through the use of selective culture media (see Section 6.3).

Remember, then, that coliforms themselves are generally not harmful. They are indicator organisms of fecal pollution which *may* contain the pathogenic organisms for diseases such as typhoid, cholera, and infectious hepatitis. Therefore, public health associations throughout North America have set standards for coliform levels that they feel are acceptable for drinking water and for water used for swimming. These experts clearly recognize the health hazard involved when sewage wastes are dumped into bodies of water used for drinking. In spite of this, 25% of U.S. municipalities still dump raw sewage into the waters. In 1970, Canada's largest city, Montreal, was dumping over 200,000,000 gallons of raw sewage per day into the St. Lawrence River! Don't shake your head in disgust until you find out what your community does with its sewage. In recent years, for example, over three-quarters of the beaches along the southern shore of Lake Erie were closed to swimming because the coliform count exceeded the safety level.

For Thought and Research

1 Consult your local public health department for information regarding acceptable total coliform and *E. coli* concentrations for drinking water and for water that is used for swimming.

2 Does your municipality treat sewage to free it of pathogenic bacteria before it is released into a natural body of water?

3 What level of government is responsible for performing coliform tests where you live? What must a citizen do to have a sample of water tested?

4 How is drinking water normally treated to insure the absence of pathogens:

    (a) at a summer cottage;

    (b) on a camping trip?

5 (a) If you live in Detroit, Windsor, Toledo, Cleveland, or any other community that uses Lake Erie as a sink, find out if your community is responsible for closed swimming beaches in the vicinity.

    (b) If you live in Buffalo, Niagara Falls, Hamilton, Toronto, Rochester, or any community that uses Lake Ontario as a sink, find out if your community is responsible for closed swimming beaches in the vicinity.

    (c) Particular cities were mentioned in (a) and (b) because serious problems exist on many of the beaches of Lakes Erie and Ontario. Regardless of where you live, does your community, at any time, raise the coliform count of local beaches to the point where swimming is prohibited?

Recommended Readings

You may wish to consult one or more introductory biology texts for more information on bacteria. Some suggested sources are:

1 *Modern Biology* by J. H. Otto and A. Towle, Holt, Rinehart & Winston, 1969.

logical Science: An Inquiry into Life, B.S.C.S. Yellow Version, Harcourt
_.ace Jovanovich, 1969.
3 *Biological Science: Molecules to Man*, B.S.C.S. Blue Version, Houghton
Mifflin, 1969.
4 *High School Biology*, B.S.C.S. Green Version, Rand McNally, 1969.

3.5   ALGAE

To the layman, algae are "pond scum," "seaweed," or "moss on
rocks in the water." To the biologist, however, algae are chloro-
phyll-bearing organisms in which each cell is capable of forming a sex cell (gamete). They are considered very primitive
plants because each cell can, by itself, carry out all life processes.
They have no specialized organs like leaves, roots, or stems.

Algae vary in size from microscopic forms to giant sea-
weeds several hundred feet long. They may be free-floating (*phytoplankton*) or they may be attached. They occur as single cells,
in clusters, and in long chains. In some cases they look very
much like "ordinary" plants (see stoneworts, Fig. 3-12). They
are found in the fresh water of lakes, rivers, streams, ponds,
swamps, puddles, and ditches; they are also found in all salt water.

Algae are classified on the basis of their methods of reproduction, the nature of their pigments, and the kinds of products they synthesize and store. This method of classification
usually results in five major divisions: blue-green algae, green algae, golden algae, red algae, and brown algae. Red and brown
algae are marine species.

Values of Algae. Algae are autotrophic; that is, they produce their own food by photosynthesis. In most aquatic ecosystems, they are the main producers and form the first link in
the food chains. They are eaten by zooplankton, which are eaten
by first-order carnivores, and so on up the food chain. As a by-
product of photosynthesis, algae produce oxygen. Some scientists
estimate that algae are responsible for up to 80% of the oxygen
production in the world. The oxygen they produce helps to purify polluted waters because it oxidizes many pollutants. It also
helps to re-oxygenate water that has had its oxygen supply lowered by pollution.

Harmful Effects of Algae. Obviously algae are a normal
and essential part of most aquatic ecosystems. Only in high concentrations do they harm the aquatic environment. After massive

algal blooms, the decaying dead algae rob the water of its oxygen. They impart undesirable taste, odor, and appearance to the water; they foul swimming beaches; they clog filters of water purification plants.

Algae as Pollution Indicators. Considerable controversy exists over the value of algae as pollution indicators. Some facts are clear though. Over 50 genera of algae can tolerate polluted water. Therefore no one species, by itself, can be used as an absolute indicator of pollution. On the other hand, certain species seem to dominate the algal blooms that choke eutrophic waters from time to time. Most biologists agree, however, that the whole algal community must be studied. The proportions of each species and the changes that occur in these proportions are better pollution indicators than any one species.

The nature of the body of water must also be considered. A cool mountain stream cannot be expected to have the same kinds of algae as a large lake. The algae found in a stream depend on such factors as current speed, depth, and mineral content. If the current is swift, only those algae which can cling to the substrate can establish themselves. Examples of such algae are *Cladophora* and *Ulothrix* (see Figure 3-12). These algae have "holdfasts" and grow in tangled chains. Planktonic algae are not found to any great extent in fast-moving streams. Why? Any that are found in such a stream probably came from a lake, pond, or backwater connected to the stream.

In slow-moving streams planktonic algae are part of the normal stream community. Even greater numbers are present in calm pools in such streams. Some algae have hair-like projections called *flagella* which enable them to swim. Such algae can maintain themselves in slow-moving waters. In the summer months slow-moving waters may be covered with algae, many of them flagellates like *Euglena* and *Chlamydomonas* (Fig. 3-12).

The abundance and types of algae present in a stream depend also on the concentration of nutrients in the water. The main inorganic nutrients which speed up algal growth are probably phosphates and nitrates (Sections 2.4 and 2.5). In an oligotrophic lake few algae are present because the nutrients required for growth are not present in large enough quantities. The few species found are usually desmids and diatoms (see Figures 3-12 and 3-16). As enrichment of the lake increases, green algae (Figs. 3-12 to 3-15) generally become more abundant. When a more advanced stage of eutrophication is reached, blue-green algae commonly predominate (Fig. 3-11).

Let us now look more closely at the types of algae you are apt to encounter during your pollution studies.

(a) Blue-green Algae (Fig. 3-11). These algae are very simple organisms. They have no nucleus; the nuclear material is scattered throughout the cell. Also, the chlorophyll is not concentrated in chloroplasts as it is in higher plants. Most blue-green algae appear pale blue under the microscope, but some appear red, purple, green, or yellow. Many species prefer eutrophic waters, including those that are polluted.

When conditions are right, blooms of blue-green algae occur in eutrophic waters. The genus *Anabaena* commonly undergoes such algal blooms. At times these blooms are so great that the water becomes discolored and the shore is littered with piles of dead algae. As they decay, foul odors are produced, making the beach unfit even for sunbathing. *Anabaena* can even undergo algal blooms in the fall when nutrient levels are usually low. This is because *Anabaena* is capable of nitrogen-fixation (see Section 2-4).

Oscillatoria is a common and widespread blue-green alga often associated with eutrophication. It grows in hair-like strands that eventually form dense clumps. Frequently these clumps have a purplish color. They are found on rocks and in pools. If you watch this alga under a microscope you will see how it got its name. Lake Washington near Seattle and Lake Zurich in Switzerland have undergone rapid eutrophication because of effluents from nearby cities. In both lakes blooms of *Oscillatoria rubescens* are common.

Other blue-green algae that are used as indicators of eutrophication are *Aphanizomenon* and *Microcystis*. These two genera along with *Anabaena* are among the dominant phytoplankton of Lake Erie. *Anacystis* often accompanies *Anabaena*. *Agmenellum*, *Phormidium*, and *Lyngbya* can also tolerate polluted water.

Rivularia is a very common blue-green alga that grows in gelatinous brown clumps on stones. It forms a thick mat which usually feels hard because of a crust of limestone. *Rivularia* is not considered an indicator of serious pollution. However, nutrients added to the water promote its growth and, when enough of the alga is present, it gives the water a musty smell. It often clogs the filters in water purification plants.

When they are present in sufficient quantities, many blue-green algae are toxic to humans and other animals. Of the toxic algae, *Anabaena*, *Microcystis*, and *Anacystis* are the most common and the most dangerous. There have been countless cases in Canada and the United States of humans who developed gastro-intestinal disorders after drinking water containing these algae.

Fig. 3-11
Some common blue-green algae.

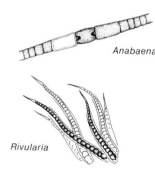
Anabaena

Rivularia

Oscillatoria

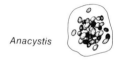
Anacystis

Lyngbya

Agmenellum

Microcystis

Aphanizomenon

Phormidium

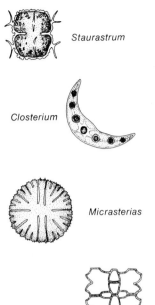

Staurastrum

Closterium

Micrasterias

Desmidium

Cosmarium

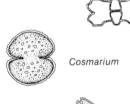

Tetmemorus

Fig. 3-12
Representatives of a group of green algae called desmids.

Fig. 3-13
Some common green algae.

Spirogyra

(b) Green Algae. All green algae have organized nuclei. Also, their chlorophyll is confined to chloroplasts. Most are grass green or yellow green. They occur as single cells, in pairs, and in colonies which are usually in the form of long threads. The plants of one group, the stoneworts, are about a foot high and look much like ordinary plants. Sometimes stoneworts are so dense on the bottom of a pond that they fill it at the rate of an inch per year! Green algae are, in general, more abundant in ponds and small lakes than all of the other algal groups combined.

One group of green algae, the *desmids* (Fig. 3-12), are characteristic of oligotrophic waters. They may, of course, be found in eutrophic waters. But if most of the phytoplankton in a body of water consists of desmids, you can be reasonably sure that the water is oligotrophic. Desmids usually occur in pairs or in filamentous colonies. The members of a pair have either a small constriction or a dividing line where they join. Desmids are usually grass green and have many interesting shapes and forms. When they are present in large numbers, they give a lake a green coloration. They prefer soft water which has a low *p*H. *Staurastrum*, *Closterium*, *Micrasterias*, *Desmidium*, *Cosmarium*, and *Tetmemorus* are illustrated in Figure 3-12. Of these genera, *Staurastrum* is commonly used as an indicator of oligotrophic conditions.

Many green algae respond well to organic pollution. One of these, *Chlorella*, is widespread throughout the United States and Canada (Fig. 3-13). It is found in nearly all waters that contain organic matter. *Chlorella* gives the water an undesirable musty odor. *Spirogyra* is also very tolerant of organic pollution. This filamentous alga is the familiar scum that often covers the surface of ponds in the early spring. It spends the winter at the bottom of the pond where it is the chief food source for many crustaceans and insects. What makes it rise to the surface in the spring?

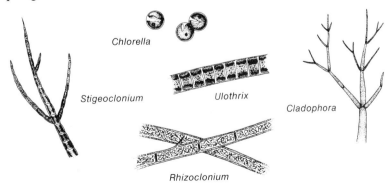

Chlorella

Stigeoclonium

Ulothrix

Cladophora

Rhizoclonium

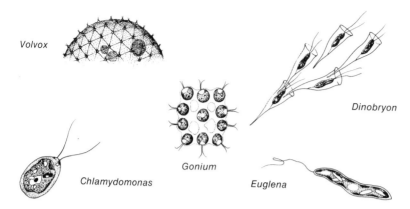

Volvox

Dinobryon

Gonium

Chlamydomonas

Euglena

Fig. 3-14
Some common flagellates.

Three other green algae commonly found in clean water are *Ulothrix*, *Cladophora*, and *Rhizoclonium*. The first two are common inhabitants of fast streams because they have "hold-fasts" that attach them to rocks. *Ulothrix* often appears as a hair-like green covering on rocks in cool streams. Only some species of *Cladophora* grow attached to rocks. Others are often rolled into balls several inches in diameter by fast currents. These balls are often seen floating down a stream. *Cladophora* also grows well in fast streams that receive sewage effluent. It is usually found far downstream from the effluent site where it appears as a stringy brown mass attached to rocks.

Some flagellates (Fig. 3-14) can be considered under the title "green algae." Sometimes they are classified as protozoa (one-celled animals) because they move (using a whip-like motion of the flagella that protrude from each cell). Generally, they are classified as protists, along with other simple organisms such as algae and protozoa. We have included them with the green algae because most species contain chlorophyll and all are single-celled. Regardless of the classification, some flagellates are important in pollution studies. *Euglena* is easily identified by its bright green color and red "eyespot." It is a common resident of slow streams and rivers. It is very common in ponds that are rich in organic matter, and thrives in organically polluted water. Even barnyard pools contain *Euglena*. It is sometimes so dense in the scum on these pools that the water appears red. Normally *Euglena* imparts a green color to water. In sunny weather, though, red pigments are often more abundant than green pigments in *Euglena*. Why is this so?

Chlamydomonas is another green flagellate that likes organically polluted water. *Dinobryon*, on the other hand, is used as an index organism of oligotrophic water. *Volvox* is a genus that you will likely find in most ponds. It is easily identified be-

Chara vulgaris

Nitella flexilis

cause it forms spherical colonies of up to 10,000 cells. *Gonium* is also found in most ponds and lakes. It forms flat plates of cells on rocks.

The stoneworts, mentioned earlier, are so much larger than other freshwater algae that they are often mistaken for higher plants (Fig. 3-15). They prefer hard water with its accompanying high *p*H. Often they form dense mats on the bottoms of ponds and shallow lakes. Limestone often precipitates from the hard water onto these plants, and collects in little lumps—hence the name stoneworts. The two common genera, *Chara* and *Nitella*, usually have a garlic-like odor.

(c) Golden Algae (Fig. 3-16). In these plants, yellow and brown pigments are more abundant than chlorophylls. The *diatoms* are the most interesting and important members of this group of algae. They are one-celled and have cell walls made of silica (the main component of sand). Each cell is composed of two halves that overlap and are held together by a band. They may occur as single cells, in pairs, or in colonies. They may be free-floating or attached to rocks and aquatic plants. When they die, they leave a golden layer of *diatomaceous earth* on the stream bed. This substance is used as an insulating material, as a filtration medium for the purification of water, and as a fine abrasive.

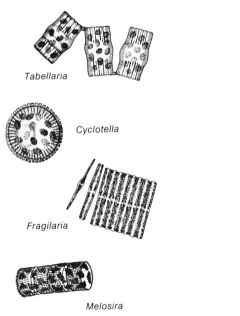

Tabellaria

Cyclotella

Fragilaria

Melosira

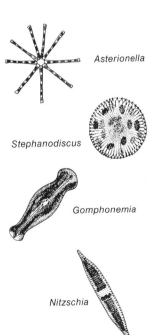

Asterionella

Stephanodiscus

Gomphonemia

Nitzschia

Fig. 3-16
Some common golden algae (diatoms).

Two species of diatoms that are characteristic of oligotrophic waters are *Tabellaria* and *Cyclotella*. If *Fragilaria*, *Melosira*, *Asterionella*, and *Stephanodiscus* are present in high concentrations, eutrophic conditions have likely arrived. *Gomphonemia* and *Nitzschia* are also tolerant of eutrophic waters.

For Thought and Research

1 Explain why each of these statements is true:

(a) The type of algae present in polluted water depends on the nature of the effluent and its concentration.

(b) If an effluent is not too concentrated, it may just increase the numbers of each of the algal species present.

(c) If an effluent is very concentrated, it will likely decrease the number of algal species present.

(d) Algae are most useful as pollution indicators if the whole algal community is studied.

(e) The proportions of the different species of algae in a body of water are more useful pollution indicators than the actual species themselves.

2 Figure 3-17 shows the effects of sewage on the total algal population in a slow-moving stream.

(a) Account for the shape of the curve.

(b) What would happen to the total algal population if the D.O. concentration were reduced to zero just below the effluent site? Why?

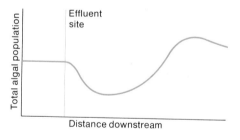

Fig. 3-17
Effect of sewage on total algal population.

3 As an oligotrophic lake becomes eutrophic, a succession of algae occurs. In general, the succession proceeds this way:

desmids → diatoms → other green algae and flagellates → blue-green algae

Collect evidence to support this generalization.

4 Find out from your local water purification plant which algae most frequently clog the filters of the plant. How do the plant operators deal with this problem?

5 What effects do you think pollutants will eventually have on the algae in the oceans of the world? How will this affect the biosphere (world ecosystem)?

6 Perform the laboratory exercise in Section 6.5.

Recommended Readings

Consult the following books for the identity of other algae.

1 *Pond Life* by G. K. Reid, Golden Press, 1967.
2 *The New Field Book of Freshwater Life* by E. B. Klots, G. P. Putnam's Sons, 1966.
3 *A Guide to the Study of Fresh-Water Biology* by J. G. Needham and P. R. Needham, Holden-Day, 1962.
4 *Algae in Water Supplies*, U.S. Public Health Service Publication No. 657.
5 *The Biological Aspects of Water Pollution* by C. G. Wilber, C. C. Thomas, 1969. See pp. 237-238.

3.6 ZOOPLANKTON

Zooplankton are microscopic and near-microscopic aquatic animals. They play key roles in most aquatic food chains. Some are scavengers which feed on dead organic matter; others are consumers feeding on algae and bacteria; still others feed on smaller zooplankton. Zooplankton, in turn, become food for higher order consumers such as mayfly nymphs and caddisfly larvae.

The use of zooplankton as pollution indicators is not yet well-established. You will, however, find them in just about every sample of natural or polluted water that you examine under a microscope. We will consider three groups of zooplankton: protozoans, rotifers, and crustaceans.

(a) Protozoans. These are, by far, the most abundant zooplankton in water. They are microscopic, single-celled organisms classified as animals by some biologists and as protists by others. They feed mainly on algae, bacteria, and particles of dead organic matter.

The members of one group of protozoa, the Sarcodina (Fig. 3-18), move by the use of pseudopods ("false feet"). The protoplasm simply flows in the direction of motion. The movement is slow but, as the organism moves over the bottom of a pond, it also feeds. The flowing protoplasm engulfs particles of organic material and bacteria. The presence of Sarcodina, there-

Fig. 3-18
The Sarcodina are common protozoans found among the zooplankton that feed on organic material.

Euglypha

Amoeba

Pelomyxa

fore, normally indicates the presence of organic matter. The source of the organic matter can, of course, be natural. The familiar *Amoeba* is a member of this group. Two other Sarcodina, *Euglypha* and *Pelomyxa*, are fairly reliable sewage indicators.

Another group of protozoa, the ciliates (Fig. 3-19), bear numerous hair-like projections called cilia. These are used for movement and for sweeping food into the animals' "mouths." Since ciliates eat dead organic matter, bacteria, and other protozoa, they are also commonly found near decaying organic matter. They are probably most abundant in the stagnant pools along the edges of ponds. *Paramecium* is the best-known ciliate. Others that you may encounter in your studies are *Coleps*, *Didinium*, *Lacrymaria*, *Colpoda*, *Colpidium*, *Glaucoma*, and *Stentor*. Some genera like *Spirostomum* are easily visible to the naked eye. *Spirostomum* is usually located near the surface of calm, well-shaded water and gives the water a white appearance. *Vorticella*, a colonial species, often appears as white patches on rocks and on aquatic animals such as turtles. Although most ciliates are found near organic matter, *Carchesium* is one of the few that is considered a sewage indicator.

(b) Rotifers (Fig. 3-20). These small unsegmented animals are often mistaken for single-celled organisms. Actually they are many-celled animals and are not related to the protozoans. Most species are barely visible to the naked eye. They are generally found only in relatively fresh water where they are the major part of the zooplankton. Their absence, therefore, could indicate pollution.

Many species of rotifers have cilia that appear to rotate like wheels as they sweep food into their mouths. They eat protozoans and algae. You will find rotifers, therefore, where you find

Fig. 3-19
Note the common features of these protozoans.

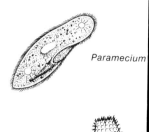
Paramecium

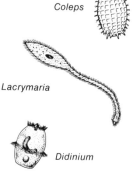
Coleps

Lacrymaria

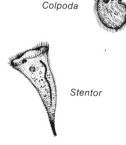
Didinium

Colpoda

Stentor

Rotaria

Philodina

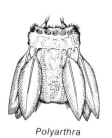
Polyarthra

Fig. 3-20
Representative rotifers.

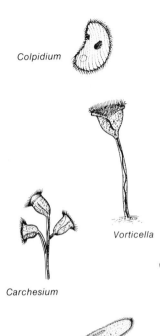

Colpidium

Vorticella

Carchesium

Spirostomum

the most protozoans and algae—at the edges of ponds. Many species attach themselves to plants and are easy to capture. Other species are free-swimming and can only be captured with a plankton net. *Philodina*, *Rotaria*, and *Polyarthra* are three common genera of rotifers.

   (c) Crustaceans (Fig. 3-21). These segmented animals are, on the average, a little larger than rotifers. You have likely seen their large cousin, the crayfish. Some crustaceans eat rotifers; others eat algae, bacteria, and protozoans; still others feed on dead organic material. Most of them require fairly high oxygen concentrations. Therefore their disappearance from a body of water could indicate pollution.

   Daphnia is the most common member of a group of crustaceans called cladocerans. Because of their appearance, cladocerans are often called water fleas. Another common genus of cladocerans is *Bosmina*. One species, *Bosmina coregoni*, is believed to indicate oligotrophic water; another species, *Bosmina longirostris* is an indicator of eutrophic water. The former is currently abundant in Lake Superior and the latter in Lake Erie. The eutrophication of Lake Zurich in Switzerland was accompanied by a shift from a high population of *B. coregoni* to *B. longirostris*.

   Cypridopsis is a common member of a group of crustaceans called ostracods or seed shrimps. These small crustaceans are found just above the bottom of ponds where they feed on bacteria and algae. They can tolerate fairly low D.O. levels.

Glaucoma

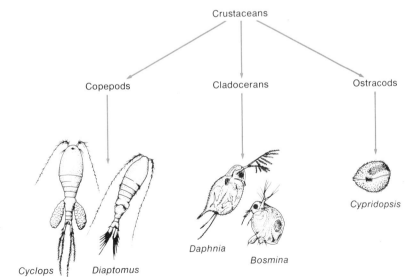

Fig. 3-21
Some common planktonic crustaceans.

The final group of crustaceans is the copepods. *Cyclops,* a common copepod, is thought to indicate pollution if it is present in unusually high numbers. *Diaptomus,* a common copepod in Lakes Michigan and Erie, is also believed to be a pollution indicator.

Summary. Don't put too much faith in any of the zooplankton as pollution indicators. The species thought to be pollution indicators may just be accidental intruders which happened to move into the water at about the same time that eutrophication began. In any case, the effect of pollution on zooplankton is usually only a secondary one. Pollution does not kill most zooplankton directly. It either lowers the D.O. level below that which the zooplankton can tolerate or it destroys their source of food. You should, therefore, only consider zooplankton in your pollution studies along with the algae, bacteria, and chemical characteristics of the water.

For Thought and Research

1 In an unpolluted body of water, zooplankton numbers are controlled largely by three factors: food supply, oxygen concentration, and predators. Trace the chain of events that might take place if a body of water becomes polluted with organic material. Pay particular attention to the effects on the zooplankton.
2 Interpret the graph in Figure 3-22.

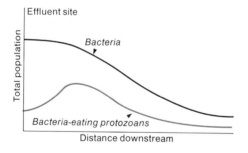

Fig. 3-22
Relationship between bacteria population and the population of protozoan predators in organically polluted water.

Recommended Readings

Use the following books to assist you in the identification of any zooplankton not illustrated in Section 3.6:
1 *Pond Life* by G. K. Reid, Golden Press, 1967.
2 *The New Field Book of Freshwater Life* by E. B. Klots, G. P. Putnam's Sons, 1966.
3 *A Guide to the Study of Fresh-Water Biology* by J. G. Needham and P. R. Needham, Holden-Day, 1962.

3.7 FISH

The types of fish in a body of water depend, to some degree, on the stage of eutrophication (Fig. 3-23). In general, the dominant fishes in oligotrophic lakes are lake trout, whitefish, walleye, lake herring, and char. As eutrophication sets in, these species remain, but in decreased numbers. Perch, black bass, pike, and

Fig. 3-23
Fish as indicators of eutrophication.

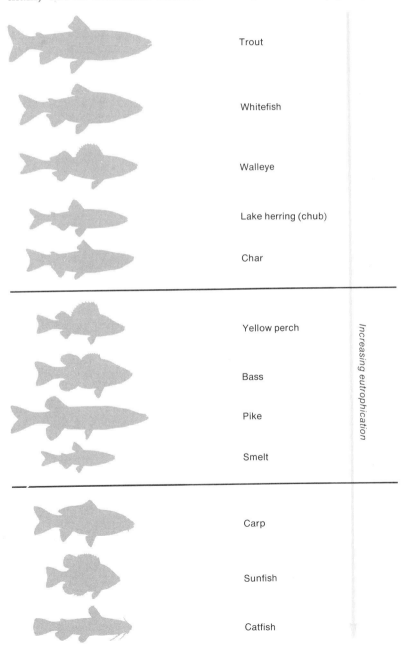

Trout

Whitefish

Walleye

Lake herring (chub)

Char

Yellow perch

Bass

Pike

Smelt

Carp

Sunfish

Catfish

Increasing eutrophication

smelt become the dominant fishes. After further eutrophication, carp, sunfish, and catfish dominate the fish population. The shift is from tasty, sports-type fish to coarse species that are not very tasty. Why does this shift occur?

The presence of perch, carp, and sunfish does not mean, however, that the water is polluted. It only means that the water is eutrophic. More specifically, it means that the D.O. concentration in deep regions is low. Pollution contributes to the problem by speeding up eutrophication.

Severe pollution of just about any kind kills most fishes. For example, 0.03 ppm of chlorine or 0.5 ppm of DDT is lethal to most trout. An effluent loaded with organic matter can lower the D.O. concentration to the point where no fish can survive. Also, some fish species eat pieces of the decaying organic matter and are poisoned by it. Pollution may also reduce the number of bottom fauna, the chief food source for many fish species. Thermal pollution can also be a factor in determining fish populations (see Section 2.8).

Consider for a moment the fish in a stream. The presence of bass or sunfish in a stream does not mean that it is polluted. It may never have contained any other species because it is shallow and warm. In fact, bass are very intolerant of pollution even though they are tolerant of warm water. If, on the other hand, you find a cool, deep stream that contains mainly trout above a town and mainly sunfish below the town, you can be suspicious of pollution. You must consider the past history of a stream as well as its present condition.

Do not rely too heavily on fish as pollution indicators. It is difficult to get reliable results because fish move so rapidly from place to place. The information that you can gather in a one-day study is not too valid. Nevertheless, a few sweeps of a seine net will give you some idea of the types of fish present. This information is helpful in judging the quality of a body of water, when it is used in conjunction with other biological tests and with physical and chemical data.

Recommended Readings

Consult these books for help in identifying fish that are not included here:
1   *A Guide to the Study of Fresh-Water Biology* by J. G. Needham and P. R. Needham, Holden-Day, 1962.
2   *Pond Life* by G. K. Reid, Golden Press, 1967.
3   *The New Field Book of Freshwater Life* by E. B. Klots, G. P. Putnam's Sons, 1966.
4   *Fishes* by H. S. Zim and H. H. Shoemaker, Golden Press, 1955.

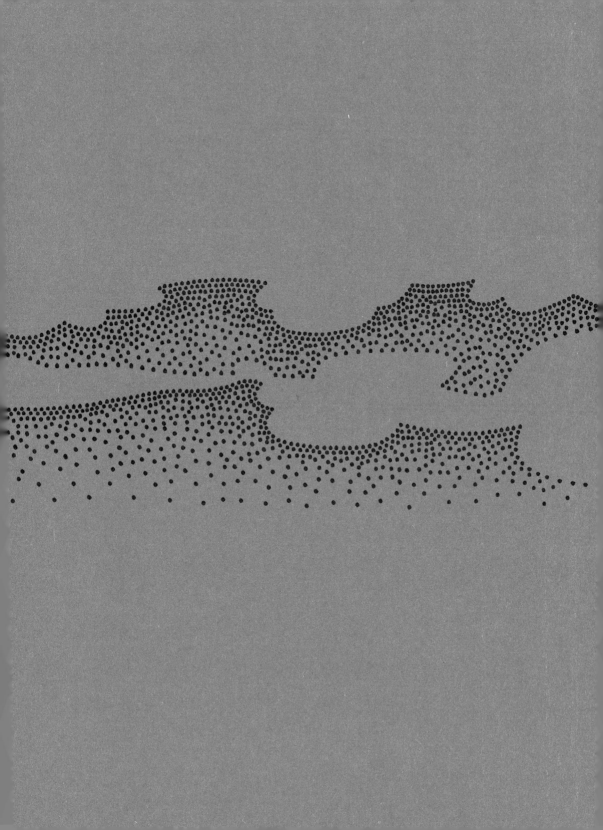

Air Pollution

4

4.1 OUR AMAZING ATMOSPHERE

Air governs the quality of our environment and the character of the world as we perceive it. Try to imagine a world without air. You would never see dawn steal across the sky or watch a bird in flight. There would be no winds, clouds, or rain. Fire would not exist because burning requires oxygen. You would live in a world of virtual silence. No form of plant or animal life could exist, even in the seas. You would no longer be protected from harmful solar radiation. Without the atmosphere, temperature fluctuations on earth would be similar to those on the moon. The temperature would soar to approximately 110°C (230°F) in the daytime and drop to –184°C (–300°F) at dark.

It is sometimes difficult to consider air a material substance. Not only is air invisible, but it also has no definite shape, volume, or density. About 5.8 thousand billion tons of air surround the earth. As the altitude increases the density of the air decreases rapidly. At a height of six miles, you could not obtain enough oxygen to live. Twelve miles up, there is insufficient oxygen to maintain a burning candle. All oxygen-dependent life is sustained by a very thin slice of the 560 mile deep atmosphere (Fig. 4-1). If we constructed a model of the earth two inches in diameter, this critical zone would be thinner than the paper of this page. Scientists estimate that 95% of all life on earth is supported by an atmospheric layer less than two miles thick.

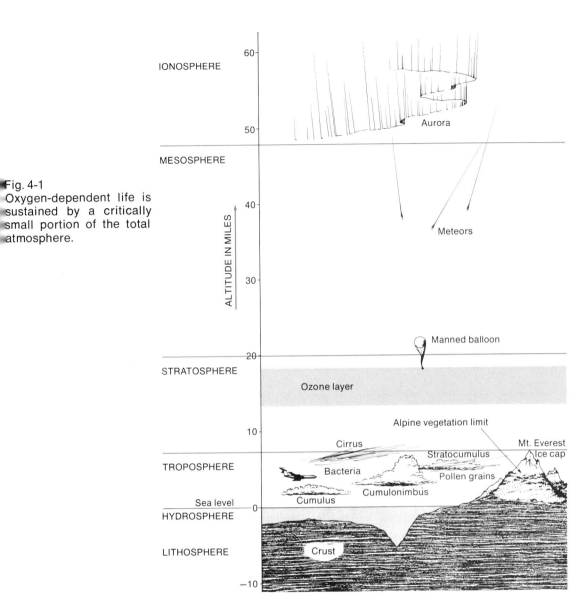

Fig. 4-1
Oxygen-dependent life is
sustained by a critically
small portion of the total
atmosphere.

An Atmospheric Anatomy. Air is a complex mixture of gases as shown in Table 8 (page 92).

For the first eight gases in the table, the composition of the air is relatively constant over the surface of the earth. Methane, ammonia, hydrogen sulfide, carbon monoxide, and nitrous oxide are present at all points on the earth's surface, but the proportion of each gas varies widely from place to place. For example, the air over a large marsh usually contains higher than average concentrations of methane, ammonia, and hydrogen sulfide. Why?

TABLE 8 THE COMPOSITION OF CLEAN, DRY AIR

Gas	Percent by volume
Nitrogen (N_2)	78%
Oxygen (O_2)	21%
Argon (Ar)	0.94%
Carbon dioxide (CO_2)	0.03%
Helium (He)	0.01%
Neon (Ne)	0.01%
Xenon (Xe)	0.01%
Krypton (Kr)	0.01%
Methane (CH_4)	trace
Ammonia (NH_3)	trace
Hydrogen sulfide (H_2S)	trace
Carbon monoxide (CO)	trace
Nitrous oxide (N_2O)	trace

The amount of water vapor also varies greatly above different regions of the earth. It ranges from 0.01% to about 4% by volume. Naturally occurring air is never entirely dry, not even above the deserts. Where would you expect to find the greatest concentrations of water vapor? If all of the atmospheric water vapor fell in one single deluge, the earth would receive a layer of water only one inch deep. Yet water vapor is one of the most crucial components of air. The vertical and horizontal movement of water vapor provides one of the earth's most important mechanisms of heat transfer. By absorbing heat from the earth's surface, this vapor maintains a life-sustaining temperature balance.

The air was never perfectly clean, even before man entered the scene. Foreign matter such as volcanic ash, bacteria, pollen, spores, salt particles from the oceans, and even cosmic dust can be found far above the earth. These particles serve a useful function. They act as nuclei around which water molecules condense to form tiny water droplets. These produce clouds and, eventually, precipitation. Problems arise only when abnormally large concentrations of foreign particles are pumped into the atmosphere, upsetting the normal cycles. Let us see how this happens.

A Balanced Biosphere. How much oxygen do you consume in a single day, in a single hour, or, for that matter, in a single minute? Since oxygen constitutes only 21% of the air by volume, how much air do you personally require in a 24-hour period to provide the necessary oxygen?

You must share the available oxygen supply with more than 3.6 billion people and with almost every other living organism on the earth. The only exceptions are the relatively few anaerobic species which can exist without oxygen.

Contrary to popular belief, green plants do not "breathe in" carbon dioxide and "breathe out" oxygen. Plants require a 24-hour oxygen supply for respiration just as animals do. In that case, why haven't we used up all of the oxygen in the air? The answer lies in the unique abilities of green, chlorophyll-containing plants, and in the constant recycling of the elements essential to life. That tiny blade of grass you tread upon has the remarkable ability to harness a portion of the sun's energy. It uses this energy to combine carbon dioxide molecules from the atmosphere and water molecules from the soil or the immediate surroundings into carbohydrate molecules. In so doing, green plants provide stored food energy which is utilized either directly or indirectly by all forms of life. This amazing process, photosynthesis, was discussed in greater detail in Section 1.4. You may recall that oxygen is released as one of the end products of this reaction:

Carbon Dioxide + Water + Light Energy

$$\xrightarrow{\text{chlorophyll}} \text{Glucose + Oxygen}$$

$$6\ CO_2 + 6\ H_2O + \text{Light Energy} \xrightarrow{\text{chlorophyll}} C_6H_{12}O_6 + 6\ O_2$$

During the daylight hours a healthy plant releases far more oxygen than it requires for its own respiration. This can be demonstrated using the set-up in Figure 4-2.

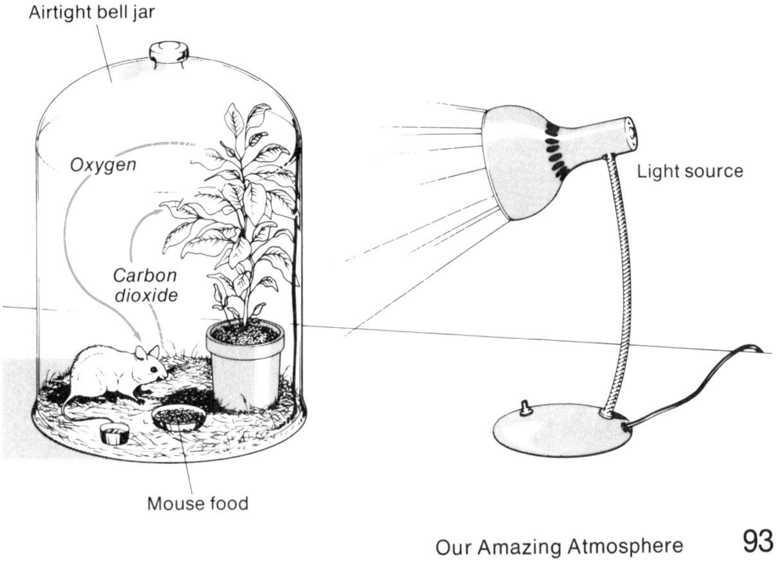

Fig. 4-2
The green plant releases enough oxygen during photosynthesis to support the respiration requirements of the mouse. The mouse exhales carbon dioxide, used for photosynthesis.

Airtight bell jar

Oxygen

Carbon dioxide

Light source

Mouse food

On a sunny day, the leaves of a full grown maple tree use about 75 cubic feet (9.2 pounds) of carbon dioxide and release an equivalent volume of oxygen. Similarly one acre of lawn uses about 900 cubic feet (111 pounds) of carbon dioxide, releasing the same volume of oxygen. As you can see, you depend upon green plants for life-giving oxygen as well as for nutrition. Yet you and all other chlorophyll-lacking organisms play an important role in the balanced cycle of nature. Each day, in the process of respiration, you convert 20 cubic feet (1.8 pounds) of oxygen to an equal volume of carbon dioxide which is then used in plant photosynthesis. Unfortunately, man, in his pursuit of progress, is now using more than his share of oxygen and is contributing excessive amounts of carbon dioxide to the atmosphere. If you live in an industrialized city, you indirectly consume 20 times your basic requirement of oxygen by the burning of fuels such as coal, oil, gas, diesel oil, and gasoline. For example, the combustion of six gallons of oil on a winter day provides heat for four people and, in the process, converts 430 cubic feet of oxygen into carbon dioxide. This means that each person requires an additional 107.5 cubic feet of oxygen just to keep warm while he is busy converting 20 cubic feet of oxygen into carbon dioxide through respiration! A 19-acre park produces enough oxygen to balance the respiration requirements of 1,000 people during the summer months. But what happens during the winter when many trees shed their leaves and discontinue photosynthesis? During this period we rely upon the oxygen in air and also upon the contributions from plant life many thousands of miles away. For example, prevailing winds over the Atlantic carry inland the oxygen produced by countless numbers of surface phytoplankton in the ocean. Inland cities such as Chicago, Toronto, and Detroit receive over 60% of their available free oxygen from ocean phytoplankton. By establishing a balance between animal and plant life, nature maintains an equilibrium between oxygen-producing and oxygen-consuming activities. What happens when man, in ever-increasing numbers, upsets this balance?

Consider the total oxygen requirement of a community of 1,000 people. To balance the oxygen–carbon dioxide cycle during an average year, including the winter months, a park of 440 acres is needed. A community of 15,000 people covers an average of 1,600 acres of land during urban and suburban development; it requires a greenbelt of over 6,000 acres to maintain an atmospheric balance. But is this the way that North American cities and towns are planned? As growing cities mushroom in all directions, countless thousands of acres of forest, grassland, and farm land are buried under buildings and a carpet of pavement.

Man is destroying the equilibrium with no thought for the possible consequences. In recent years, scientists have measured a decrease in the amount of atmospheric oxygen. Correspondingly, the concentration of atmospheric carbon dioxide has risen. Oxygen constitutes a relatively large proportion of the total atmosphere. Presumably we are in no danger of exhausting our supply in the near future. Our greatest problem could well be the increasing concentration of carbon dioxide.

For Thought and Research

1 How much air do you personally require during a 24-hour period? Determine the number of inhalations which you make during this time interval and measure the volume of air involved each time. The volume exhaled in one breath can be measured by the displacement of water from a large bottle.

2 What percentage of the total urban land area is developed as park land in the town or city nearest your home? Compare this figure with those of large cities around the continent.

3 How many acres of rural land were used for urban development during each of the previous 10 years in your province or state? Use graphical methods to estimate future requirements for this purpose, and comment on the validity of your estimates.

4 (a) How many miles of new pavement were laid last year in your state or province? You can obtain this figure from your Department of Highways. Measure the width of an average highway. Calculate the approximate number of acres of land that were covered with cement or asphalt last year.

(b) Obtain from your Department of Highways the number of miles of pavement laid during each of the previous 10 years. Plot a graph of number of acres covered versus the year and extend the graph to estimate the acreage that will be covered 10 years from now. How valid is your estimate? What factors might affect it?

Recommended Readings

1 *Investigating the Earth*, Earth Science Curriculum Project, Houghton Mifflin, 1967. This book contains interesting readings and investigations regarding the atmosphere.

2 *The Earth* by A. Beiser and the Editors of *Life*, Life Nature Library, Time, Inc., 1963. This colorful book has an excellent section on the structure of the atmosphere.

3 *Weather* by P. D. Thompson, R. O'Brien, and the Editors of *Life*, Life Science Library, Time, Inc., 1965. Consult this book to find out how the atmosphere controls heat and oxygen movement over North America.

4 "The Biosphere," *Scientific American*, September, 1970. This article explains the grand-scale cyclic mechanisms of life on earth.

4.2 AIR POLLUTION—ITS NATURE AND CAUSES

Air pollution is defined as the presence in the atmosphere of substances or radiations which adversely affect living organisms or their habitats. Air pollution has basically the same causes as water pollution. Impurities are introduced in such abundance that they cannot be adequately absorbed or removed before they accumulate in harmful concentrations. Many of these impurities are of natural origin. Natural pollutants include airborne pollen, gases produced by decay, and dust from erosion. To these, man adds his own harmful contaminants at an ever-increasing rate. Unhampered, nature insures a continued supply of the basic elements required by living organisms. Molecules of natural wastes are recycled for future generations. However, the air is not self-purifying with respect to high concentrations of unnatural wastes. After being discharged into the air, these pollutants are diluted by air currents and may be chemically converted to other forms. They either remain indefinitely airborne or, after precipitation from the atmosphere, contribute to soil and water pollution.

Sources of Air Pollutants. These fall into three major categories: surface friction, vaporization, and combustion.

The force of friction is constantly at work, wearing down exposed surfaces. This process is a major source of airborne solid particles of various sizes. Consider the numerous activities associated with the average construction site. The sawing, drilling, and grinding of materials such as asphalt, steel, and lumber contribute vast quantities of particulate matter to the air. Can you suggest any areas in your own neighborhood where you would expect to find a high concentration of airborne particles contributed by frictional sources?

Vaporization is the physical change of a liquid to the gaseous state. Vapors are also produced directly from the solid state by a physical change called sublimation. A volatile liquid or solid is one which readily produces vapors. Many vapor pollutants result from the use of highly volatile liquids, such as the solvents in fast-drying paints and adhesives. Other vapors are emitted during chemical reactions involving liquids or solids under conditions of high temperature and low pressure. Industries such as those which produce metals, chemicals, paints, and rubber are most commonly associated with vapor by-products. Visible fumes often result from the condensation of such vapors. Are there any industrial processes in your area which involve vapor production? If so, is an attempt being made to control vapor emission into the atmosphere?

Centuries ago, man learned to depend upon fire. Today, as never before, man relies upon the process of combustion to support the technological world which he has developed. Combustion is the rapid chemical union of a substance with oxygen. The reaction liberates heat and, often, light energy. The products of combustion are oxides, as the following examples illustrate:

$$\text{Carbon} + \text{Oxygen} \rightarrow \text{Carbon Dioxide} + \text{Energy}$$
$$C + O_2 \rightarrow CO_2 + \text{Energy}$$

$$\text{Hydrogen} + \text{Oxygen} \rightarrow \text{Hydrogen Oxide (Water)} + \text{Energy}$$
$$2\,H_2 + O_2 \rightarrow 2\,H_2O + \text{Energy}$$

$$\text{Sulfur} + \text{Oxygen} \rightarrow \text{Sulfur Dioxide} + \text{Energy}$$
$$S + O_2 \rightarrow SO_2 + \text{Energy}$$

Consider the chemical composition of the most commonly used fuels. Coal and its derivative, coke, are chiefly carbon. Petroleum and natural gas are composed of organic compounds called hydrocarbons. As the name suggests, a hydrocarbon is a molecule containing carbon and hydrogen atoms. Petroleum consists of many different hydrocarbons, some of which are very large, complex molecules. By refining petroleum, different mixtures of hydrocarbons are obtained. Each mixture provides a certain grade of fuel which has a specific use (Fig. 4-3).

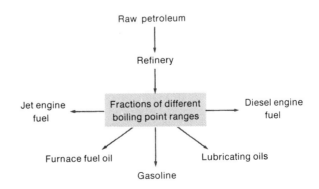

Fig. 4-3
Each grade of fuel refined from petroleum has a specific use.

When a hydrocarbon burns completely, the carbon and hydrogen combine with oxygen to produce carbon dioxide and water respectively.

Hydrocarbon + Oxygen → Carbon Dioxide + Water + Energy

The rapid expansion of the hot gaseous products drives the pistons of countless engines. The amount of oxygen consumed in any combustion process depends upon both the quantity and chemical composition of the fuel which is burned. Another determining factor is the extent to which the molecules of the fuel effectively mix with oxygen molecules during the reaction.

A single pound of carbon requires 11.5 pounds of air for complete combustion. A single pound of hydrogen gas requires 34.5 pounds of air and a single pound of sulfur, 4.3 pounds of air for complete combustion. Most fuels contain impurities such as sulfur and nitrogen compounds. These molecules also react with oxygen during combustion. As you can see, the amount of air required for a complete reaction will vary considerably depending upon the fuel involved.

In actual practice, combustion is rarely complete. How is combustion affected if an insufficient amount of air is available, or if the fuel and oxygen molecules do not mix properly? A number of different products are obtained in such cases. Some of the fuel molecules will react completely, others will only partially combine with oxygen, and the rest will remain unreacted. This type of reaction is termed *incomplete combustion*. It results in many of the troublesome pollutants associated with the burning of fuels. For example, the incomplete combustion of carbon produces unburned carbon particles (soot), carbon dioxide, and the deadly gas, carbon monoxide. In contrast, complete combustion produces only carbon dioxide.

The Automobile—Pollution on Wheels. The products just mentioned are, of course, components of automobile exhaust. The combustion of gasoline in automobile engines accounts for more than 60% of the total annual air pollutant emission in North America. Cars are mobile sources of a host of contaminants: carbon monoxide, incompletely burned hydrocarbons, nitrogen oxides, sulfur oxides, solid particles, and compounds of elements such as lead and phosphorus. These latter compounds are contained in fuel additives and lubricating oils. Many factors determine the quantity of pollutants which an individual car contributes. The design of the engine and the grade of gasoline which it requires are important elements. Combustion is far more efficient when the car is driven at a constant speed with a minimum of stopping and starting. To obtain the necessary power, a car burns a greater amount of fuel while accelerating. An engine designed to run with maximum efficiency at high speeds is highly inefficient while the motor is idling. The most critical factor affecting the efficiency of an engine is the maintenance which it receives. Modern, high-powered engines require regular tuning to insure the most efficient conversion of fuel to energy. One important aspect is the adjustment of the engine for the best possible mixture of fuel and air during combustion. Otherwise, an excessively large portion of the gasoline is not properly burned, but is emitted as undesirable waste products.

California, faced with critical smog problems, initiated legislative controls designed to reduce automobile pollutants. These were later adopted nationwide, and have become more stringent. The earlier controls restricted exhaust emissions of carbon monoxide and hydrocarbons. The controls on the 1971 models reduced the high vaporization rate of hydrocarbons from carburetors and fuel systems. California also limited emission of nitrogen oxides from 1971 model cars and small trucks.

How effective are such controls? One study, involving vehicles in normal use, investigated 300 passenger cars in five cities. Results indicated that the cars equipped with control devices emitted about half as much hydrocarbon waste and carbon monoxide as the cars without controls (Fig. 4-4). The efficiency

Fig. 4-4
The projected effects of the motor vehicle emission control program in Los Angeles County. (Data published by Air Pollution Control District, Los Angeles County.)

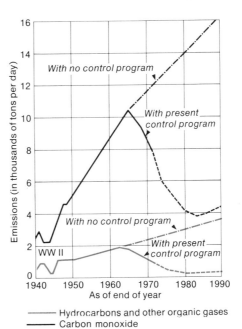

Hydrocarbons and other organic gases
Carbon monoxide

of such control devices diminishes somewhat as car mileage increases, but new and improved control systems are being produced. These devices must have a greater testing period before their long-term effectiveness can be properly assessed. However, such advances provide one bright side to an otherwise gloomy situation. The problem now facing car manufacturers is to gain public acceptance and cooperation for the proper maintenance of these control devices. At best, such pollution controls may only be temporary. Manufacturers never expect to produce an internal combustion engine which emits absolutely no pollutants. Further, by 1975, the increased numbers of automobiles producing emissions are expected to counteract the effectiveness of any proposed control devices. How can this dilemma be solved? Are electric cars the answer? Or will the constant recharging of such battery-operated vehicles simply contribute other forms of pollution? Eventually, North Americans may have to adjust their entire way of life to insure their continued existence.

For Thought and Research

1 List the major friction sources of air pollutants in your area.

2 Why are motorists requested to shut off car engines while the gas tanks are being filled at service stations? What is the average quantity of gasoline spilled during a single refueling process? What other sources of vapor pollutants are found in your area? What types of vapor emission controls are used?

3 To demonstrate the products of complete and incomplete combustion, ignite different combinations of acetylene gas (a hydrocarbon) and air. To produce the acetylene (C_2H_2), react calcium carbide (CaC_2) with water.

(a) Invert a medium-sized test tube filled with water into a 400 ml beaker containing about 200 ml of water. Using forceps, drop a small lump of calcium carbide into the water. Place the inverted test tube over the calcium carbide and allow the gas produced to completely displace the water in the test tube. Keeping the test tube inverted at all times, remove it from the beaker and place it mouth-down on a stable surface.

(b) Repeat this procedure with three more test tubes filled with water, varying the amounts of gas collected: displace one-half, one-third, and approximately one-twelfth of the water in each of the three consecutive tubes. As you remove each tube from the beaker, place your thumb over its mouth after allowing air to replace the remaining water in the tube. Shake the test tube up and down to mix the air and acetylene. Then place the tube mouth-down until you are ready to ignite the mixture.

(c) Hold each test tube in a nearly horizontal position and apply a burning splint to its mouth. Make a note of all observations. After each reaction is complete, add 4 ml of limewater to the test tube and shake. Observe any change in the limewater.

(d) Place some fresh limewater in a clean test tube. Using a piece of tubing or a straw, bubble carbon dioxide from your breath through it. How does the change in the limewater compare with the change you observed in part (c)? Suggest the products which were formed during combustion in each test tube. Which reactions would you classify as complete combustion? Which as incomplete combustion?

4 (a) Would it be feasible to limit the size of automobiles driven in congested city areas? To what extent would such a policy reduce air pollution? Are less powerful engines a better solution?

(b) Investigate the advantages and the disadvantages of electric cars with respect to air pollution. (See *Recommended Reading* 9.)

5 (a) Discuss the role of the automobile in the North American economy.

(b) Investigate other problems concerning automobiles such as consumption of oil reserves; consumption of various raw materials, for example, iron and nickel; loss of agricultural land for highway construction; noise pollution; and traffic congestion in large cities.

Recommended Readings

1 "Our Ecological Crisis," *National Geographic*, December, 1970.

2 "Some Burning Questions About Combustion," *The Environment* by the Editors of *Fortune*, Harper & Row, 1970.

3 *Vanishing Air* by J. C. Esposito, Ralph Nader's Study Report on Pollution, Grossman Publishers, 1970. Your best source of legal details regarding air pollution.

4 "The Darkening Veil," *Man Against His Environment* by R. Rienow and L. T. Rienow, Ballentine Books, 1970.

5 "Nature and Sources of Air Pollution," *The Pollution Reader* by A. DeVos et al., Harvest House, 1968.

6 "The Air We Breathe," *Pollution Probe* by D. A. Chant, New Press, 1970.

7 *Cleaning Our Environment. The Chemical Basis for Action*, American Chemical Society, 1969. See the section "Controlling Air Pollution" for more detailed information about automobile standards.

8 "The Control of Air Pollution" by A. J. Haagen-Smit, *Scientific American*, January, 1964. An excellent account of automobile emission controls.

9 "Menace in the Skies," *Time*, January 27, 1967. See the section dealing with electric car research.

10 *California Test Procedure and Criteria for Motor Vehicle Exhaust Emission Control*, State of California Motor Vehicle Pollution Control Board, 1969 (revised). This manual explains the testing of car emissions.

4.3 AIRBORNE PARTICLES

Have you ever noticed the tiny moving particles revealed by a beam of sunlight? Conscientious housekeepers engage in a constant battle against the "fallout" of such particles from the air. But the most critical surface that receives them is the human lung. As you read this page you are probably breathing at a rate of 14 to 18 times per minute. In a large city such as Montreal,

New York, Chicago, or Toronto, each breath carries about 70,000 solid particles into your lungs. Even a person enjoying "clean" country air inhales about 40,000 particles in each breath. Do these figures sound unbelievable? How are they determined? What is the nature and the source of these particles? And, most important, what is their effect, not only on the lungs, but also on every other surface which they encounter?

Sources of Solid Particles. Airborne particles represent the most obvious and also the most complex type of pollutant. The chimneys of the world pour out countless tons of soot daily which are distributed throughout the atmosphere by the prevailing winds (Fig. 4-5). Particles generated over a city normally remain in the air for only a few days, although the lighter ones may drift for weeks. Larger types, such as fly ash and soil, usually settle close to their source. The lifetime of a particle in the air also depends upon the height at which it is released. Large or intense sources such as nuclear explosions, forest fires, volcanic eruptions, and concentrated industries can produce particles which travel thousands of miles.

The unit most commonly used to measure airborne particles is the micron. It is represented by the symbol μ and is equal to 10^{-4} cm or 39.4 millionths of an inch. The largest airborne particles are more than 10 microns in diameter. They are produced largely by mechanical processes in nature such as grinding, spraying, and erosion. Examples of erosion at work are the dust storms in the American southwest. Photographs of these storms vividly illustrate the infiltration of airborne soil even into closed buildings.

Most particles in the air range from 1 to 10 microns in diameter. They result from natural processes and from the wastes produced by an ever-growing population. These wastes include carbon and soot, fly ash, grease, oil, and metal fragments. Each automobile contributes, in the form of atmospheric particles, a measurable percentage of the weight of gasoline burned. In addition, minute portions of rubber are worn off the surface of spinning tires. One survey estimates that 50 tons of rubber particles fall daily on the streets of Los Angeles alone!

Products of combustion and condensation tend to predominate in any sample of particles ranging in diameter from 0.1 to 1 micron. Particles smaller than this are obviously difficult to study. They are attributed to combustion processes since concentrations of this size of particle are much greater in city air than elsewhere. A unit volume of air above downtown Toronto contains three times the weight of these particles as an equal volume of air tested in a rural center. These smaller particles tend to

Fig. 4-5
Chimneys throughout the world belch countless tons of waste into the atmosphere.

remain suspended indefinitely. They contribute greatly to the problem of haze.

Measurement of Airborne Particles. Filtering methods are most commonly used to measure the amount of particulate matter in the air. A filter is selected which has pores small enough to trap the particle size being studied. To measure all of the solids in an air sample, an extremely fine filter must be used. After pumping a measured amount of air through the filter, the residue is removed, weighed, and analyzed. Very small amounts of air must be tested at any one time, since the filter holes tend to clog quickly. Viewed through a microscope, many of the captured particles can be identified according to size, shape, color, or crystal structure. Chemical testing identifies others.

A second method employs settling to trap the particles. Open containers or glass slides coated with a sticky material are placed in strategic locations for varying time intervals. The collected fallout is later weighed and analyzed. This simple procedure is used to measure the amount of fallout per day, week, or month.

Considering the various factors in the production and distribution of these particles, would you expect a fairly constant deposition during the course of the year in your area? What factors affect the rate of deposition? In cities across North America, seasonal changes in temperature influence the use of heating and cooling systems. How does this affect production of particulate matter? Is generation of electricity greatest during the day or during the hours of darkness? Does the direction of the prevailing winds change greatly at different times of the year? How does this affect the eventual distribution of airborne particles? Are the majority of industries more active at certain times of the day or of the week? Does the volume of traffic fluctuate measurably during the daily or the weekly period? As you can see, solid particle pollution involves many complicated factors which must be carefully studied before steps can be taken successfully to improve conditions.

Common Properties of Airborne Particles. Before scientists can consider the effects of airborne particles, they must first determine their physical and chemical properties. All of these particles share many physical properties including growth by condensation; adsorption (the attraction of vapors and gases to the particle surface); absorption (the penetration of vapors and gases into the intermolecular spaces of the particle); coagulation (the grouping together of similar particles); dispersion (the scattering or spreading out of originally concentrated particle groups); and the ability to absorb or to scatter light.

Generalizations about the chemical behavior of such particles are impossible because of the diversity of material involved. Since the smaller airborne particles collide frequently, chemical interactions must occur, but studies in this area are greatly hindered by particle size.

Effects of Airborne Particles. The immediate effects of fine particles in the air are all too familiar. Airborne dirt is constantly settling on our surroundings. This steady influx also affects the breathing processes of living organisms. The mucous lining and the cilia of your nose and throat usually catch the larger particles. The smaller particles are believed to penetrate the deeper recesses of the lungs. Blackened lung tissue of city dwellers indicates that at least a portion of this inhalation is retained. Although the actual soot is not yet positively related to specific diseases, the chemicals adsorbed by the soot are of serious concern. Soot consists mainly of carbon. Carbon particles are used in gas masks because they can adsorb large quantities of certain gases. Thus inhalation of airborne soot may permit a more rapid and concentrated penetration of dangerous gases than would normally occur. Scientists at the Pasadena (California) Foundation for Medical Research filtered samples of city air. The collected particles were added to groups of living cells in a human tissue culture. A number of these cells started to enlarge; some divided with unusual rapidity; others divided into abnormal shapes. This behavior is typical of cancer cells. The scientists concluded that a number of the fine particles are apparently carcinogens (cancer-producing substances).

Similar studies are being conducted in research centers throughout the world to determine the effects of particle pollution on humans. For example, many allergies are directly connected to airborne particles such as dust and pollen. Some medical researchers suspect a correlation between the increase in soot particles and the increase in cases of pneumonia. In addition, high and low concentrations of cadmium particles apparently correlate with high and low death rates from heart disease in many North American cities. Ore smelting plants are believed responsible for the dispersal of cadmium particles.

The finer particles in the atmosphere create layers of haze which reflect artificial light reaching these layers from below. You may have observed this phenomenon when approaching a city at night. However, this haze layer may well create a far more menacing situation. The latest measurements indicate that the mean global temperature has been decreasing in recent years. Many scientists attribute this to the steadily increasing global concentration of airborne particles. These particles scatter the incoming sunlight, effectively reducing the amount reaching the surface of the earth. This tends to lower the temperature. A backward glance into history reveals a sobering example of this effect.

In 1815, a volcano named Mount Tamboro erupted on the island of Sumbawa in Indonesia. Vast quantities of ash were blasted high into the air and spread by the winds. Areas more than 380 km away received a 25 cm fallout of volcanic dust. An estimated 150 cubic kilometers of ash were spread through the atmosphere, eventually affecting climate around the world. In 1816, farmers in the northern American states waited in vain for summer to arrive. The British "summer" was the coldest on record. The average July temperature there was only 13.4°C (56°F) as compared to the usual 15.7°C (61°F). Reduced crop yields resulted in record high grain prices. One interesting observation from Britain's weather statistics reveals that the three coldest decades occurring over a 250-year interval coincide with three periods of intense volcanic activity, as Table 9 indicates.

TABLE 9

Period of low temperatures	Period of volcanic activity		
	Volcano	Location	Year of eruption
1781–1790	Mt. Asama	Japan	1783
	Mt. Skaptar	Iceland	1783
1811–1820	Mt. Tamboro	Indonesia	1815
1881–1890	Mt. Krakatoa	Java (Sundra Strait)	1883

Consider the possible consequences if airborne particles continue to be released into the atmosphere from man-made sources at the current rate. Could such increasing concentrations permanently alter the global climate and hence disrupt growing seasons and the critical world food supply? Some experts predict that, if the current trend continues, a new ice age will begin within the foreseeable future.

For Thought and Research

1 Heavy particles are those which measure more than 50 microns in diameter. A recent study of heavy particle dustfall in Toronto, Ontario, revealed the following data:

Toronto suburbs	15 tons/square mile/month
Toronto downtown area	35 tons/square mile/month
Toronto waterfront	110 tons/square mile/month

Clearly the amount of particulate matter that a Torontonian inhales depends upon his general location within the city. Suggest a reasonable interpretation of these data. (If you are not familiar with Toronto, study its location on a map of Ontario.) Would you expect these figures to be the same for all months of the year? Why? Would you expect to find a similar pattern if you performed this study in your town or city? Why?

2 Perform the study in Section 7.1 to check your prediction in question 1.

3 In one large North American city, an average suburban lot of 6,000 square feet receives from 0.3 to 0.5 pounds of dustfall daily. Make a list of the contributing factors near your school. From this list, estimate the daily fallout on the school property. Check your prediction using the procedure outlined in Section 7.1. You may also wish to perform this study on the lot on which your home or apartment is located.

4 Perform the studies outlined in Sections 7.2–7.5.

5 Obtain permission to perform the study outlined in Section 7.2 in a room filled with cigarette smoke. Compare the results with those obtained immediately outside your school and in the downtown area. What conclusions do you draw from the comparison?

6 How do you think plant growth is affected by the accumulated dustfall and by specific components of the dustfall? Design and perform suitable investigations. Don't forget to use controls.

Recommended Readings

1 *This Vital Air, This Vital Water* by T. G. Aylesworth, Rand McNally, 1968.
2 *Air and Water Pollution* by G. Leinwand, Washington Square Press, 1969.
3 *Cleaning Our Environment. The Chemical Basis for Action*, American Chemical Society, 1969.
4 *The Pollution Reader* by A. DeVos et al., Harvest House, 1968. See the appropriate parts of "Nature and Sources of Air Pollution."
5 "The Control of Air Pollution" by A. J. Haagen-Smit, *Scientific American*, January, 1964.
6 "Air Pollution" by N. Hinch, *Journal of Chemical Education*, February, 1969.

4.4 SULFUR DIOXIDE

The Origin of the Problem. In information on air pollution, reference is usually made to sulfur dioxide. You may live in an area where the sulfur dioxide level is monitored daily as part of the *"pollution index."* This clear, colorless gas and its derivatives constitute about 18% of the total accumulation of air pollutants. This is of considerable importance because sulfur dioxide, with its sharp, choking odor, is one of the most dangerous gases, particularly to man.

Sulfur dioxide is only one of several forms in which airborne sulfur circulates globally. The sulfur cycle also includes hydrogen sulfide gas, sulfuric acid aerosol, and many sulfate salts in aerosol form. (An aerosol is a suspension of liquid droplets or solid particles in a gas. For example, cigarette smoke is an aerosol of ash in air. Fog is an aerosol of water droplets in air. What is smog?) Study Figure 4-6 to determine the origin of the various sulfur compounds and the manner in which they are interrelated.

What becomes of all this airborne material? A given volume of sulfur dioxide gas entering the air will be removed as an acid or a salt by precipitation or by gravitational settling within five days to two weeks. Unfortunately, the problem does not end at this stage. The effectiveness of sulfur dioxide simply continues in another form.

Figure 4-6 shows that almost 80% of sulfur dioxide is produced from hydrogen sulfide gas originating from natural sources. Man's contribution of 20% doesn't seem like much. But man has altered the cycle of airborne sulfur to the extent that a

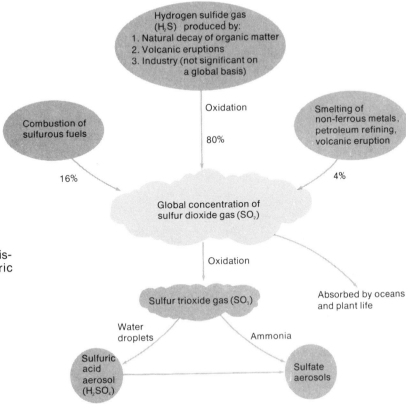

Fig. 4-6
The production and dispersal of atmospheric sulfur dioxide.

critical problem exists in many cities. Recent estimates of global emissions of sulfur dioxide indicate that 16% of the total airborne concentration is from the combustion of sulfur-containing fuels such as coal and oil. Global figures can be misleading because they suggest a uniform distribution over the entire earth. In fact, the concentration of sulfur dioxide varies greatly in different regions. Obviously the major consumers of coal and oil are concentrated in large, industrialized cities where the amount of fuel burned is proportional to the population. But the degree of sulfur dioxide pollution depends not only on the amount but also on the type of fuel burned. The various kinds of coal and oil contain differing amounts of sulfur. When coal with a high sulfur content is burned, up to 10% of its weight may enter the atmosphere as sulfur dioxide. For example, studies conducted over the past decade revealed that New York City had the highest average and maximum levels of sulfur dioxide recorded in any major American city. This uniquely bad pollution problem resulted from use of a cheap, sludgy grade of fuel oil with a high sulfur content. Under a new fuel oil code, a cleaner grade of oil must now be used. New York state led eastern North America in the establishment of rigid standards limiting the sulfur content in fuels.

Figure 4-7 shows the American national contributions of sulfur oxides produced by the combustion of fossil fuels. These emissions accumulate over a city and usually disperse slowly. Unfavorable climatic and geographic conditions as well as poor circulation of city air due to tall, congested buildings are often contributing factors. Consider some of the cities which you know to be acutely affected by air pollution, New York and Los Angeles, for example. Analyze their situation in view of the above factors. Is sulfur dioxide a problem in your own area? If so, can you suggest the causes?

Fig. 4-7
Nationwide sources of sulfur oxides emissions— U.S. 1968. (Data published by U.S. Department of Health, Education, and Welfare.)

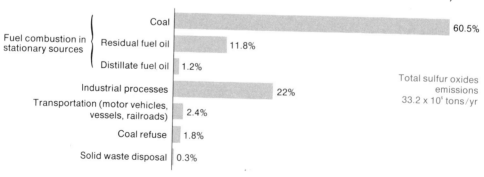

Fuel combustion in stationary sources
Coal — 60.5%
Residual fuel oil — 11.8%
Distillate fuel oil — 1.2%
Industrial processes — 22%
Transportation (motor vehicles, vessels, railroads) — 2.4%
Coal refuse — 1.8%
Solid waste disposal — 0.3%

Total sulfur oxides emissions
33.2×10^6 tons/yr

TABLE 10

Time period	Maximum concentration	
	Industrial–commercial	Residential
1-hour average	0.40 ppm	0.25 ppm
24-hour average	0.20 ppm	0.10 ppm
Annual average	0.05 ppm	0.02 ppm

Air Pollution Standards. Standards have been established in many areas for the maximum levels of sulfur dioxide permitted in residential and industrial areas. These are stated as the concentration of sulfur dioxide per one million parts of air by volume (ppm). Variation exists in the levels allowed over different periods of time. Table 10 outlines regulation standards of the Ontario Air Pollution Control Act, 1967. During the winter these standards are exceeded almost daily in Canadian cities such as Toronto and Montreal.

The effect of sulfur dioxide upon human beings is widely debated in the medical research field. Many authorities feel that the concentrations of sulfur dioxide considered "safe" by the existing standards represent a long-term health hazard. As the concentration of sulfur dioxide increases, the initial annoyance gradually leads to breathing difficulties and eventually to severe irritation. A sulfur dioxide level of 6 ppm paralyzes and corrodes the respiratory organs. The duration of the exposure is also a critical factor. The results of a five year study conducted in New York City were presented to the American Public Health Association in Philadelphia during the autumn of 1969. The data showed that deaths started to rise sharply at a level of 0.20 ppm of sulfur dioxide in the air. At the same time, a team of Buffalo doctors released a study of the number of hospitalized children under 16. Hospital admissions for asthma jumped from 32.4 per 100,000 during periods of low air pollution to 50.7 during times of high air pollution. Cases of eczema increased five-fold. While some authorities reject sulfur dioxide as the critical factor, many others feel that the present regulations allow excessively high sulfur dioxide levels.

Legislation, however, is apparently inadequate for effective enforcement, even at the existing standards. Toronto tests indicated that the levels of sulfur dioxide in the air exceeded the Government's regulations on 53 occasions during the first six months of 1968. Clearly, such laws are only as meaningful as the measures employed to enforce them. But is this possible unless the general public is fully informed about the extent and hazards of such pollution? Consider, for example, the situation in the Province of Quebec. Statistics indicate that air pollution in Montreal has reached the same dangerous level as that of Chicago, a much larger city. More than 800 tons of sulfur dioxide escape into the Montreal air daily! In the downtown area, the yearly average sulfur content in the air has soared to four times the designated "safety" level. Readings have climbed nearly as high in the residential suburbs. Yet the Quebec health officials have not published readings of provincial pollution levels, maintaining that such statistics would alarm people having an inadequate scientific background to properly assess the data. However, recent public pressure has resulted in the release of a report which has initiated proposals for new city bylaws to gradually reduce the sulfur content in the coal and fuel oil used in Montreal.

A Problem of Power. Electric power plants maintain the pulse of our modern world. Those fired by coal or oil are also the third largest source of air pollutants and the leading producer of sulfur oxides in North America. Although the increasing use of nuclear power should reduce the amount of pollutants emitted relative to the total amount of power produced, fossil fuels remain an important and growing source of air contaminants. By the year 2000, nuclear plants may well supply more than half of the electric power on this continent. However, by that time the total power required is expected to be six or seven times greater than at present. This means that at least three times as much fossil fuel will be used! The most plentiful fossil fuel reserve is coal. Although North America has substantial reserves of low-sulfur coal, most of these are located far from the major metropolitan markets. As a result, more than 90% of the coal burned in North American power plants today has a sulfur content greater than 1%. Domestic coals may contain up to 8% sulfur.

Studies are being conducted to determine the extent to which sulfur compounds can be removed from coals. However, such cleaning processes create an economic burden for the consumer. This cost may be reduced if the sulfur compounds removed prove to be valuable for any chemical processes. Unfortunately, until such problems are resolved, coal of markedly lower sulfur content is unlikely to gain widespread use. Although the petroleum industry is having more success in reducing the sulfur content, residual fuel oil accounts for less than 10% of utility power production. Some plants use low-sulfur natural gas and many others are seriously studying the feasibility of conversion to this fuel. However, natural gas is the least abundant of all the domestic fossil fuels.

Obviously the problem of eliminating sulfur oxide pollutants is a difficult one. The use of tall stacks is generally regarded as unsatisfactory; the "cleaning" of available fossil fuels involves tremendous expense. Can other more effective control measures be developed?

Effects of Sulfur Oxides. Sulfur oxides combine with moisture to form sulfurous acid and the extremely corrosive sulfuric acid.

$$\text{Sulfur Dioxide } + \text{ Water} \rightarrow \text{Sulfurous Acid}$$
$$SO_2 + H_2O \rightarrow H_2SO_3$$

$$\text{Sulfur Trioxide } + \text{ Water} \rightarrow \text{Sulfuric Acid}$$
$$SO_3 + H_2O \rightarrow H_2SO_4$$

Each day, the air which you inhale passes through the nasal cavity and the windpipe to contact directly an area 25 times greater than your exposed skin surface. This region is provided by the tiny membranes of your lungs. Every square inch of your respiratory system provides moisture, an ideal reactant for the sulfur oxides which enter your system. (They also irritate the eyes and the skin.) How is the delicate lung tissue affected? Clinical studies on humans are surprisingly limited to date. Yet even if sulfur dioxide were not regarded as a health hazard, its destructive nature would warrant the removal of this gas from the air.

Acids forming in the atmosphere fall with rain and create havoc with crops and wild plants. Since lichens store the acids in their roots and then die, botanists use them as indicators of sulfur dioxide fallout. This accurate test has designated Ontario cities such as Sudbury, Hamilton, and Toronto as "lichen deserts" because these plants cannot survive in the surrounding areas. The numerous crop plants killed, even at low fallout levels, include wheat, barley, oats, white pine, cotton, alfalfa, buckwheat, sugar beet, and a score of others. Within a 20 mile radius of Kingston, Tennessee, 90% of the white pine trees have been killed. Sulfur dioxide from a Tennessee Valley Authority power plant is the recognized culprit. Over 50 years ago, two copper smelters near Ducktown, Tennessee, released enough airborne sulfur to poison the surrounding soil. Even today, this land remains almost totally devoid of vegetation.

Concentrations of airborne sulfur compounds also threaten aquatic life. Rain and snow absorb these compounds and carry the resulting acids into soil, rivers, lakes, and ponds. Most aquatic organisms cannot survive when the *p*H falls below

4.0 (see Section 2.3). Some species of fish, such as salmon, die when the pH drops to 5.5. The pH of natural waters is affected by many factors. Seasonal changes in temperature and biological activity, as well as the mineral content of the water, must be considered. Nevertheless, the decreasing pH of many bodies of water has been directly related to increasing concentrations of airborne sulfur compounds.

Sulfurous and sulfuric acids attack virtually any exposed metal surface, such as steel rail tracks. They also react with such substances as brick, stonework, and even granite. Sulfur oxides cause discoloration and embrittlement of plastics, rubber, paper, and countless other materials. In urban areas afflicted with excessive emissions, sulfur oxides and their acid by-products etch away buildings, bridges, steel girders, automobiles, and highways. One sad piece of evidence is "Cleopatra's Needle." This giant monument survived Egypt's sunshine and abrasive sandstorms for 35 centuries before its transfer to New York City in 1880. In a short span of less than 100 years, the clean-cut hieroglyphics have been eroded away. Museums, galleries, private homes, and national monuments alike suffer from this destruction of art treasures. In our relentless quest for a bountiful future, are we destroying the immeasurable values of the past?

For Thought and Research

1 What type of fuel is used by the power plant which supplies your community? What is the average sulfur content of this fuel? What economic factors contribute to the choice of this particular fuel?

2 Have standards governing sulfur dioxide levels been established for your community? If so, how do they compare with the standards used by other centers? Do the sulfur dioxide levels in your community ever exceed these standards? If so, what type of action is taken?

3 You may wish to investigate some of the characteristics of sulfur dioxide gas in the laboratory. Use a Bunsen burner to ignite a pea-sized piece of sulfur in a deflagrating spoon. USE A FUME HOOD! CAUTIOUSLY direct a small sample of the resultant gas toward you and note the characteristic odor. Test the gas with moist litmus paper. Design a method to prove whether the gas is more or less dense than air. (Hint: The gas will decolorize dilute potassium permanganate solution.)

4 Perform investigations 7.6, 7.7, and 7.8.

Recommended Readings

1 *Cleaning Our Environment. The Chemical Basis for Action*, American Chemical Society, 1969.

2 *This Vital Air, This Vital Water* by T. G. Aylesworth, Rand McNally, 1968.

3 *The Pollution Reader* by A. DeVos et al., Harvest House, 1968. See the appropriate parts of "Nature and Sources of Air Pollution."

4 *Pollution Probe* by D. A. Chant, New Press, 1970. See the chapter on "Air Pollution."

5 Most of the *Recommended Readings* in Section 4.2 contain material on sulfur dioxide pollution.

4.5   CARBON MONOXIDE

(a) Sources. Carbon monoxide (CO), a colorless, odorless gas, accounts for more than 51% of the total annual air pollutant emissions in North America. This gas is almost entirely a man-made pollutant. The most significant source is incomplete combustion, during which each carbon atom combines with only one atom of oxygen. Estimates show that automobile engines alone contribute more than 80% of the *global* carbon monoxide emissions. Combustion in industry, power plants, residential heating, and refuse disposal accounts for the remainder (Fig. 4-8). Photochemical reactions (reactions initiated by light) of hydrocarbons in polluted atmospheres also produce tiny amounts of carbon monoxide. Very few natural sources of this gas are known. Under abnormal conditions plants can produce carbon monoxide. Certain marine organisms, such as jellyfish, can emit gas bubbles containing as much as 80% carbon monoxide. However, these natural contributions are not considered significant.

(b) Effects. Carbon monoxide is extremely dangerous. Just 10 ppm of carbon monoxide in air is sufficient to cause illness. Within 30 minutes, 1,300 ppm is fatal. Many deaths attributed to carbon monoxide have resulted from inhalation of exhaust fumes from a running automobile engine in a closed garage. Approximately 3 pounds of carbon monoxide are released

Fig. 4-8
Nationwide sources of carbon monoxide emissions—U.S. 1968. (Data published by U.S. Department of Health, Education, and Welfare.)

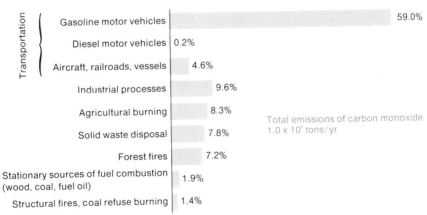

Transportation
- Gasoline motor vehicles — 59.0%
- Diesel motor vehicles — 0.2%
- Aircraft, railroads, vessels — 4.6%

Industrial processes — 9.6%

Agricultural burning — 8.3%

Solid waste disposal — 7.8%

Forest fires — 7.2%

Stationary sources of fuel combustion (wood, coal, fuel oil) — 1.9%

Structural fires, coal refuse burning — 1.4%

Total emissions of carbon monoxide
1.0×10^8 tons/yr

for every gallon of gasoline which is consumed by automobile engines. This means that about 300 pounds of carbon monoxide are released for every 100 cars on the road during an average 15 mile drive. Yet for many years, authorities claimed that carbon monoxide did not present a hazard as long as it was released into the open air. More recent scientific studies contradict this theory with increasing evidence of chronic carbon monoxide poisoning. Although carbon monoxide levels are not high enough to prove fatal, the effects of gradual poisoning may accumulate in the body. They produce mild symptoms such as drowsiness, nausea, or dizziness.

Such mild carbon monoxide poisoning is suspected as the cause of many highway accidents which were formerly blamed on driver fatigue. A one-hour exposure to a carbon monoxide concentration of 120 ppm can significantly impair driving ability. Yet measurements of this gas indicate concentrations as high as 100 ppm in tunnels, parking garages, and the streets of major cities such as New York, Chicago, Detroit, and London, England. Many people complain of headaches after driving to work engulfed in the exhaust produced by heavy traffic. Faulty car exhaust systems are particularly dangerous because carbon monoxide tends to seep up from beneath the floor and poison the occupants. In addition, the smoking of a single cigarette in an unventilated car has been shown to raise the carbon monoxide level sufficiently to affect driver and passengers alike.

The effects of carbon monoxide poisoning are similar to those of oxygen starvation. Hemoglobin in the red blood cells carries oxygen, absorbed from air in the lungs, to the rest of the body cells. Unfortunately, hemoglobin finds carbon monoxide at least 200 times more attractive than oxygen. Carbon monoxide also forms stronger and more efficient chemical bonds to hemoglobin than does oxygen. A complex molecule, carboxyhemoglobin, is formed. This complex is very stable and the important hemoglobin molecules are not released to continue oxygen distribution for several hours. During this time, the body is effectively deprived of its necessary oxygen, just as if many of the red blood cells had been lost. An 8 hour exposure to an atmosphere containing 80 ppm of carbon monoxide reduces the oxygen-carrying capacity of the blood about 15%, or more than would the loss of one pint of blood.

Ordinarily, we have a greater exposure to carbon monoxide during our active daily lives. Traffic jams can produce carbon monoxide levels as high as 400 ppm! Any physical exertion increases both the rate of respiration and the amount of air inhaled each time. Ironically, we breathe in more carbon monoxide just when our individual oxygen requirements are greatest. Obese people and those who suffer from respiratory, chronic heart, or vascular diseases are particularly affected by carbon monoxide. Heavy smokers, who regularly inhale a personal supply of carbon monoxide, may have as much as 5% of their hemoglobin permanently combined with carbon monoxide. It is not surprising to doctors that habitual smokers are among the first to require hospital treatment during severe air pollution episodes. Carbon monoxide concentrations reach the toxic level in their bloodstreams faster than in those of non-smokers.

Recently, tests were completed involving people who worked in offices located within 100 feet of main traffic arteries in New York City. Although office air was filtered through air conditioners, the carbon monoxide level was three times higher than the level which affects mental processes. When blood is unable to supply the brain with sufficient amounts of oxygen, coordination and mental agility suffer. Since the performance of these workers in a normal atmosphere had never been compared, the problem had not been recognized. For example, research volunteers reacted only 60% as accurately to sound after inhaling 50 ppm of carbon monoxide for almost an hour. A driver or pedestrian affected to this extent would certainly be more accident-prone.

The obvious remedy to this growing problem is a great reduction of traffic on city streets. In Tokyo, Japan, congested streets restrict the multitude of cars to the speed of a bicycle. Japanese policemen, directing rush hour traffic, must stop every half hour to breathe from oxygen tanks located at station houses! Any attempted traffic restrictions have swiftly proven successful. Beginning in August, 1970, desperate Tokyo authorities banned automobiles from downtown Tokyo every Sunday. Within one hour of the initial closure, the carbon monoxide level dropped from 10.5 ppm to 2.3 ppm. Delighted Japanese officials have now made the Sunday ban permanent. Surely other cities can learn from this example.

For Thought and Research

1 Present car engines release about 3 pounds of carbon monoxide for each gallon of gasoline consumed. Based on the approximate distance driven and the average mileage obtained, calculate the amount of carbon monoxide released by your family car:

(a) during an average working day;
(b) during one week;
(c) during one month;
(d) during the course of a year.

2 Apart from engine modifications, could the contribution of carbon monoxide from the car or cars used by your family be significantly reduced without causing radical changes in your family activities? Would you, your family, and your neighborhood support legislation to reduce the number and use of automobiles in order to combat increasing air pollution?

3 Is traffic congestion a problem in your area? If so, where and when would you expect to find the highest concentration of carbon monoxide? You can test your prediction using the sampling technique explained in Section 7.6.

Recommended Readings

1 "The Air We Breathe," *Pollution Probe* by D. A. Chant, New Press, 1970.
2 *The Poison Makers* by R. D. Lawrence, Nelson, 1969. See Chapter 2 on the effects of carbon monoxide.
3 "Gases in the Air," *This Vital Air, This Vital Water* by T. G. Aylesworth, Rand McNally, 1968.
4 "Menace in the Skies," *Time*, January 27, 1967.
5 "Air," *The Pollution Reader* by A. DeVos et al., Harvest House, 1968.
6 *Cleaning Our Environment. The Chemical Basis for Action*, American Chemical Society, 1969.
7 "Air Pollution," *Population, Resources, Environment* by P. R. Ehrlich and A. H. Ehrlich, W. H. Freeman, 1970.
8 "Air Pollution" by N. Hinch, *Journal of Chemical Education*, February, 1969.

4.6 CARBON DIOXIDE

Carbon dioxide is a natural atmospheric component which does not significantly react with other airborne substances. Hence you may wonder why this gas is included in a unit on air pollution. But if you refer back to Sections 4.1 and 4.2 the reason should become apparent. The combustion of fossil fuels in increasing quantities has raised the global concentration of carbon dioxide beyond its natural level.

You encountered the carbon cycle in Section 2.2. One vital portion of this cycle involves the exchange of carbon dioxide between the air and living organisms. This exchange occurs fairly rapidly and, as a result, has little effect on the global atmospheric concentration of the gas. Another portion of the carbon cycle includes the oceans. They provide a giant "sink" for carbon dioxide. The exchange occurs in two phases. First, the carbon dioxide is exchanged between the atmosphere and the upper 50 to 100 meters of water. A second exchange takes place between this surface layer of water and the deeper water beneath. The extent of the initial phase depends on such properties of the surface waters as acidity, temperature, and salt content. In some regions, such as the tropics, carbon dioxide is released by the seas instead of dissolving as is usually the case.

On a global scale this overall balancing mechanism is extremely slow. Whenever a vast amount of carbon dioxide is suddenly released into the atmosphere, the balance of the carbon cycle is upset. For example, an estimated 520,000 planes fly in and out of New York City yearly. They contribute about *36 million tons* of carbon dioxide to the city atmosphere. Approximately 5 years elapse before the carbon dioxide can be gradually absorbed by the ocean surfaces of the world to readjust the relative proportions of carbon dioxide in the atmosphere and in the seas. Another 1,500 years could pass before a balanced exchange of this new concentration of carbon dioxide was established between the upper and lower levels of the oceans.

Another natural process which removes atmospheric carbon dioxide occurs in the oceans. As the sea water absorbs more carbon dioxide, it becomes increasingly acidic and reacts with the limestone (calcium carbonate) in the ocean sediments. The reaction ties up carbon dioxide in the form of bicarbonate (see Section 2.2). However, this process would require at least 10,000 years to readjust the atmospheric content of carbon dioxide.

Carbon dioxide is also removed from the air by the tremendously slow weathering of silicate rocks. This process forms limestone ($CaCO_3$) and dolomite ($CaCO_3 \cdot MgCO_3$).

By burning fossil fuels, man is releasing vast quantities of carbon dioxide into the air. He is doing this much faster than these fuels originally formed from sediments of organic matter. Meanwhile, millions of square miles of vegetation which absorb carbon dioxide for photosynthesis are being replaced with cities and highways. The carbon cycle is out of balance. If the oceans could absorb carbon dioxide as fast as it is produced, each new emission would be distributed as follows: 83% would end up in the ocean "sink," leaving only 17% of the emission in the atmosphere. At present, however, only 50% of the carbon dioxide emission is actually being absorbed by the oceans. The remainder is accumulating in the atmosphere at a rate of 6,000,000 tons yearly. Nature's slow removal processes simply cannot keep pace with the rate at which carbon dioxide is being produced by fossil fuel combustion.

Carbon dioxide is the only combustion product whose increase has been documented around the globe. Apparently the concentration of atmospheric carbon dioxide has increased since 1860 from approximately 290 ppm to about 320 ppm—an increase of more than 10%. Precise measurements have determined that the carbon dioxide content increased by 6 ppm between 1958 and 1968 alone. Projected estimates predict a 25% increase beyond the 1970 concentration to about 400 ppm by 2000 and to between 500 and 540 ppm by 2020. Since carbon dioxide is not directly harmful to living organisms, why are scientists so concerned about this increase?

Effects. Atmospheric carbon dioxide produces a phenomenon called the "greenhouse effect." Carbon dioxide molecules absorb energy in the infrared region of the spectrum. This includes most of the thermal or heat energy which normally radiates back into space from the surface of the earth. Carbon dioxide absorbs this escaping heat radiation, trapping it, much like a blanket of insulation. As early as 1899, an American geologist warned that growing concentrations of atmospheric carbon dioxide could produce an increase in global temperatures. At the present rate of carbon dioxide emission, trapped heat could raise the temperature of the earth's surface by 22 C° (40 F°) within 500 years. Even half of this increase would result in violent circulation of air masses and storms destructive beyond the imagination. Many scientists fear that average temperatures may rise enough within decades to melt the polar ice caps! Ocean levels would then rise more than 100 feet, drowning the coastal cities of the world.

Ironically, airborne particles and aerosol emissions could more than counteract this effect by acting as a screen which reduces the amount of sunlight reaching the surface of the earth. Theoretically, a 25% increase in atmospheric turbidity would offset a 100% increase in the carbon dioxide concentration. In addition, higher temperatures increase surface water evaporation and, hence, cloud formation. Clouds, in turn, reflect incoming sunlight before it reaches the earth's surface.

The remaining forest regions may also influence future conditions. Increased atmospheric carbon dioxide can stimulate more rapid plant growth. The increased photosynthesis should then remove more carbon dioxide from the air. Eventually this carbon dioxide will be returned to the atmosphere when the plants die and decay. However, the forests, which account for almost two-thirds of land photosynthesis, have a long life span and thus effectively retain carbon dioxide. The complexity of the situation defies reliable prediction. However, one fact is certain. The great disturbances now under investigation are being induced by man himself. And eventually this tampering with biological and geochemical balances may well prove disastrous to life as we know it on earth.

1 As a result of plant photosynthesis, the land areas of the earth convert about 40 billion tons of atmospheric carbon dioxide into larger carbon compounds each year. The rate at which carbon dioxide is consumed varies greatly over the surface of the earth. Explain how each of the following factors would affect the amount of carbon dioxide used by plant life:

    (a) the geographical region of the earth;
    (b) the type of vegetation (for example, tropical rain forest, tundra);
    (c) seasonal changes in climate.

2 The forests of the world represent the main consumers of carbon dioxide on the land surface. If you measured the carbon dioxide concentration at various levels above a forest floor, you would detect changes during a 24-hour period. The average concentration of carbon dioxide is approximately 320 ppm. Changes in biological activity alternately decrease this level by 10 to 15 ppm or increase it beyond 400 ppm.

    (a) Explain the changes in the concentration of carbon dioxide during a 24 hour period.

    (b) When would you expect to measure the maximum concentration of carbon dioxide; the minimum concentration of carbon dioxide? Explain your answers.

    (c) At which level above the forest floor would you expect to find the greatest fluctuations in concentration? Why?

    (d) During which season would forests of the northern hemisphere have the greatest reducing effect on the atmospheric carbon dioxide concentration? Why?

3 The amount of carbon dioxide consumed annually by phytoplankton in the oceans is about 40 billion tons (roughly equivalent to the total intake of carbon dioxide by land vegetation).

    (a) What is the major source of this carbon dioxide?
    (b) What are other sources of carbon dioxide in the marine environment?
    (c) What factors determine the amount and rate of carbon dioxide consumption by these tiny plants?

Recommended Readings

1 *Cleaning Our Environment. The Chemical Basis for Action*, American Chemical Society, 1969.

2 *Scientific American*, September, 1970. See the articles titled "The Carbon Cycle" and "Human Energy Production as a Process in the Biosphere." Both give excellent presentations of the role of carbon dioxide.

3 *This Vital Air, This Vital Water* by T. G. Aylesworth, Rand McNally, 1968. See the section entitled "Radioactive and Miscellaneous Pollutants in the Air."

4.7   NITROGEN OXIDES

From your study of the nitrogen cycle in Section 2.4 you can appreciate the important role of nitrogen and its many compounds in the balance of nature. Nitrogen gas (N_2) comprises 78% of the total atmospheric volume. Only a few organisms, cosmic radiation, and lightning are able to "fix" or combine atmospheric nitrogen with other elements to form the nitrogen compounds essential for plant and animal growth. The global nitrogen cycle is extremely complex and many of the interrelationships have not yet been fully studied. Man has disrupted many of nature's cycles, but one of his greatest interventions has been the industrial fixation of nitrogen to produce fertilizers. Man now converts as much nitrogen into fertilizer compounds *each year* as did all of the huge terrestrial ecosystems before the dawn of modern agriculture. This appears to be an important advance over nature's relatively slow conversion processes. However, until now, the nitrogen removed from the air by natural fixation processes was returned at almost the same rate by organisms which released nitrogen from these compounds. What will happen over an extended period of time if nitrogen is removed from the atmosphere much faster than it can be returned?

To date, man's contributions of nitrogen compounds to the atmosphere are considered relatively insignificant on a global scale. However, nitrogen oxides now represent more than 10% of the annual air pollutant emissions. And, considering the effects, a little goes a long way.

(a) Sources. Nitrogen atoms react with oxygen atoms to produce eight possible combinations or molecules called nitrogen oxides. Only three of these—nitrous oxide (N_2O), nitric oxide (NO), and nitrogen dioxide (NO_2)—are found in significant concentrations in the atmosphere. Furthermore, only the latter two are involved in air pollution problems.

Nitrous oxide (N_2O), the most abundant of the three, is a colorless gas which is relatively unreactive toward other atmospheric substances such as ozone, oxygen, and hydrocarbons. The global concentration of approximately 0.25 ppm N_2O is apparently produced entirely from natural sources.

Atmospheric nitric oxide (NO) is a recent discovery. Limited data on non-urban Panama air provided the first indication that nitric oxide might now be a trace constituent of the atmosphere. If so, the increasing amount of combustion could be a major source. Combustion, particularly at high temperatures and pressures, converts atmospheric nitrogen chiefly to nitric oxide. As Figure 4-9 indicates, this nitric oxide is oxidized, either slowly by oxygen or rapidly by ozone, to produce nitrogen dioxide.

Nitrogen dioxide (NO_2) is a toxic reddish-brown gas with the acrid odor characteristic of nitric acid. Man-made sources do *not* account for a major portion of the low global concentration

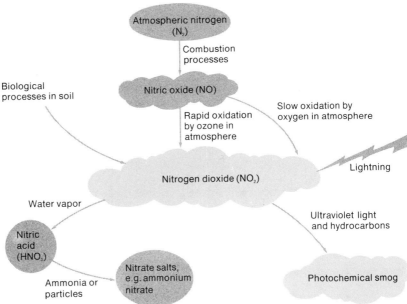

Fig. 4-9
The production and dispersal of atmospheric nitrogen dioxide.

of this gas. Nevertheless the effects of nitrogen dioxide are becoming increasingly important in polluted atmospheres.

In 1970, approximately 23 million tons of nitrogen oxides were emitted in the United States alone. Nitrogen in the air forms varying amounts of nitrogen oxides in all internal combustion engines and fuel furnaces. Motor vehicles and stationary sources burning coal, oil, and natural gas are the major contributors (Fig. 4-10). Concentrations of nitrogen oxides are

Fig. 4-10
Nationwide sources of nitrogen oxides emissions—U.S. 1968. (Data published by U.S. Department of Health, Education, and Welfare.)

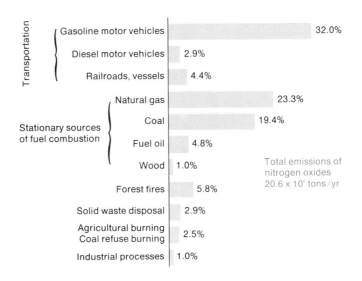

closely related to population densities. As expected, over 60% of the total emissions occur in urban areas.

Nitrogen oxides are also important by-products in certain chemical processes. Examples are the manufacture of sulfuric acid using the lead chamber process, nitric acid production, and the synthesis of nylon intermediates.

Nitric oxide (NO) is the main nitrogen oxide in engine exhaust. It forms in the cylinders when nitrogen and oxygen combine during combustion. Greater amounts of nitric oxide form at higher combustion temperatures. For this reason some of the engine adjustments proposed to reduce hydrocarbon and carbon monoxide emissions tend to increase nitric oxide production. Thus control of nitrogen oxides is a complicating factor in the battle to balance emission control against engine performance. In California, all 1971 model cars and light trucks were equipped with control systems to limit the release of nitrogen oxides. Although the technology required is complex, this control should soon be adopted throughout North America.

(b) Effects. Both nitric oxide and nitrogen dioxide have damaging effects on man and his environment. Nitric oxide (NO), similar to carbon monoxide (CO), reduces the oxygen-carrying capacity of the blood. It also combines readily with oxygen to yield nitrogen dioxide (NO_2). When nitrogen dioxide contacts water vapor in the air or body, the highly corrosive nitric acid (HNO_3) forms. Hence nitrogen dioxide irritates the eyes, nose, bronchial tubes, and lungs. High concentrations of this toxic gas prove fatal.

Nitrogen dioxide, a strong oxidizing agent, attacks metals and other materials rapidly. Nitrogen dioxide has also been identified as the "trigger" for the photochemical reactions which produce the type of smog most commonly associated with Los Angeles. The nitrogen dioxide molecule can absorb ultraviolet energy from the sun. Each energized molecule then initiates a complex series of reactions with atmospheric hydrocarbons. (See Section 4.13.) The prospect has been raised that the future reduction of hydrocarbons by emission controls on automobiles will only intensify the effectiveness of unreacted nitrogen dioxide in the air.

For Thought and Research

1   The Hearn Generating Station supplies 10% of the electrical power for the province of Ontario. In an effort to eliminate the sulfur dioxide emissions from burning sulfur-containing coal, the station is converting to natural gas. Referring to Figure 4-10, predict how this conversion will affect nitrogen oxide emissions. Do you think the conversion will improve air quality?

2 (a) You may wish to investigate the properties of nitrogen dioxide. WORKING UNDER A FUME HOOD, carefully drop a small piece of copper into a small amount of concentrated nitric acid (HNO_3) in a beaker. The following reaction takes place:

$$\text{Copper} + \text{Nitric Acid} \rightarrow \text{Copper(II) Nitrate} + \text{Nitrogen Dioxide} + \text{Water}$$
$$Cu + 4\ HNO_3 \rightarrow Cu(NO_3)_2 + 2\ NO_2 + 2\ H_2O$$

Observe the resultant gas. Be careful not to inhale any of this toxic vapor.

(b) The effects of this gas on various materials can be investigated using the procedure outlined in Section 7.9.

3 What are the major sources of nitrogen oxides in your area?

4 Carbon monoxide and unburned hydrocarbons result from the incomplete combustion of gasoline. The state of California initiated engine modifications in order to reduce these emissions. Referring to Figure 4-11, explain how the attempted reduction of carbon monoxide (CO) and hydrocarbon (HC) emissions has affected the emissions of nitrogen oxides. Why is the concentration of nitrogen dioxide a critical problem in Los Angeles?

Fig. 4-11
The projected effects of motor vehicle emission controls in Los Angeles County. (Data published by Air Pollution Control District, Los Angeles County.)

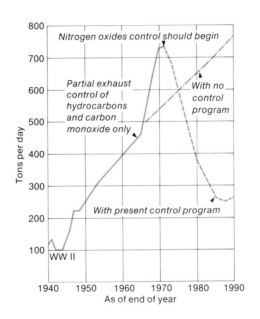

Recommended Readings

1 *Cleaning Our Environment. The Chemical Basis for Action,* American Chemical Society, 1969. See the sections "Nitrogen Oxides" and "Control Technology: Nitrogen Oxides" for a more detailed description of emission control problems.

2 *Scientific American,* September, 1970. See "The Nitrogen Cycle" and "Human Energy Production as a Process in the Biosphere."

3 *Scientific American*, January, 1964. The article titled "The Control of Air Pollution" outlines the role of nitrogen dioxide in photochemical smog production and the control of nitrogen oxides from stationary sources as well as from motor vehicles.

4 *Nationwide Inventory of Air Pollutant Emissions—1968*, U.S. Department of Health, Education, and Welfare, National Air Pollution Control Administration, Raleigh, N.C., August, 1970. This publication provides data on nitrogen oxides and other major pollutants.

4.8 HYDROCARBONS

(a) Sources. Hydrocarbons are molecules composed of only two types of atoms—hydrogen (H) and carbon (C). Yet they exist in an almost infinite variety of sizes and shapes. For example, about 200 different kinds of hydrocarbons are emitted from every automobile exhaust. Natural sources of hydrocarbons include forests, vegetation, forest fires, and the bacterial decomposition of organic matter. The latter produces vast amounts of methane, CH_4.

Man-made emissions account for only 15% of the total atmospheric content of hydrocarbons. Unfortunately, these emissions are concentrated in urban areas and produce pollution problems. They also provide the reactants for photochemical smog production. Man-made sources of hydrocarbons include industrial processes, organic solvent evaporation, and incineration. The combustion of a single ton of coal releases almost 20 pounds of hydrocarbons. However, the major contribution results from the processing and use of petroleum (Fig. 4-12).

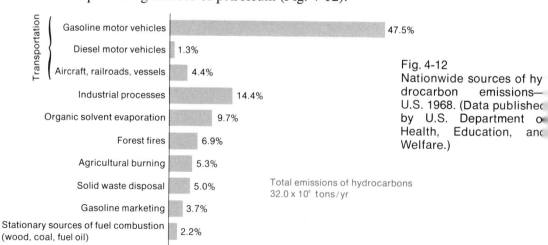

Transportation
Gasoline motor vehicles — 47.5%
Diesel motor vehicles — 1.3%
Aircraft, railroads, vessels — 4.4%
Industrial processes — 14.4%
Organic solvent evaporation — 9.7%
Forest fires — 6.9%
Agricultural burning — 5.3%
Solid waste disposal — 5.0%
Gasoline marketing — 3.7%
Stationary sources of fuel combustion (wood, coal, fuel oil) — 2.2%
Structural fires, coal refuse burning — 0.9%

Fig. 4-12
Nationwide sources of hydrocarbon emissions—U.S. 1968. (Data published by U.S. Department of Health, Education, and Welfare.)

Total emissions of hydrocarbons
32.0×10^6 tons/yr

The handling and marketing of gasoline involves significant evaporation during the filling of tank trucks and automobile tanks. Gasoline also vaporizes from the automobile carburetor and fuel system. In addition, internal combustion engines exhaust unburned or incompletely burned hydrocarbons. The emission of hydrocarbons by diesel engines is negligible, despite the nauseating odor produced.

(b) Effects. Apart from the role of hydrocarbons in photochemical smog, some may present an independent danger. Many hydrocarbons are carcinogenic—they can cause cancer. For example, the hydrocarbon benzopyrene is one of the compounds present in cigarette smoke. World-wide studies have reported that benzopyrene produces lung cancer. Most city residents—infants, children, and adults—inhale as much benzopyrene from the air as they would from seven cigarettes a day. But if you find this figure disturbing, just consider the cities in Table 11. In which city would you expect to find the highest rate of lung cancer?

TABLE 11

City	Benzopyrene level equivalent (cigarettes per day)
Birmingham, Ala.	50
Cincinnati, Ohio	26
Detroit, Mich.	37
Nashville, Tenn.	40
New York, N.Y.	45

Most airborne benzopyrene results from the combustion of coal. About 10% is emitted from automobile exhaust, some evaporates from road and roof tar, and some is contributed by oil fires and processed rubber. The level of this hydrocarbon varies greatly from city to city. And benzopyrene is only one type of carcinogenic hydrocarbon; at least five others are known to exist in polluted air.

The level of any hydrocarbon found in the urban atmosphere is not entirely related to population density. For example, hydrocarbon concentrations are generally higher in warmer weather. Why? The chemical stability of most hydrocarbons is greatly dependent upon the weather conditions. Industrialized European cities have considerably higher and more seasonally variable hydrocarbon levels than the large American cities. Why?

Any direct association between many types of airborne hydrocarbons and their suspected effects is difficult to prove. But, hydrocarbons represent more than 13% of the yearly air pollutant emissions and controls must be designed for them.

For Thought and Research

1 Referring to Figure 4-12, determine the major sources of hydrocarbon emissions in your area. How could these emissions be significantly reduced? How would you expect the concentration of hydrocarbons in the atmosphere to vary during the seasons in your area?

2 Is your family car equipped with hydrocarbon emission control devices? If it is, find out how they work. Does the owner of the car have such control devices checked regularly to avoid mechanical deficiencies in their operation? Would you support legislation compelling people to install and maintain such control devices?

3 The following data comparing emissions from gasoline-powered and diesel-powered motor vehicles were published in the Nationwide Inventory of Air Pollutant Emissions, 1968.

TABLE 12 NATIONWIDE EMISSIONS
(Millions of tons per year)

Pollutant source	Motor vehicles	Gasoline	Diesel
Carbon monoxide	59.2	59.0	0.2
Hydrocarbons	15.6	15.2	0.4
Nitrogen oxides	7.2	6.6	0.6
Particulates	0.8	0.5	0.3
Sulfur oxides	0.3	0.2	0.1

Diesel-powered vehicles make up less than 1% of the vehicle population. They consume about 4% of the vehicular fuel.

(a) Which pollutants are emitted to a lesser extent by a diesel engine than by a gasoline engine of comparable power?

(b) If all vehicles were converted to diesel power, what would be the most noticeable change in air quality?

(c) You have likely driven behind a diesel bus or truck. What were its most noticeable pollutants? Which of these emissions can be controlled better than they are now? (Consult *Recommended Reading* 1.)

(d) If the controls suggested in (c) were enforced, would conversion of all vehicles to diesel power lessen air pollution? Would such a conversion be economically feasible?

Recommended Readings

1 *Cleaning Our Environment. The Chemical Basis for Action*, American Chemical Society, 1969. See "Motor Vehicles: Diesel."
2 *Nationwide Inventory of Air Pollutant Emissions—1968*, U.S. Department of Health, Education, and Welfare, Raleigh, N.C., August, 1970. This publication provides data on hydrocarbons and other major pollutants.
3 "Air Pollution" by N. Hinch, *Journal of Chemical Education*, February, 1969.
4 "The Control of Air Pollution" by A. J. Haagen-Smit, *Scientific American*, January, 1964. This article gives an excellent explanation of the exhaust devices which are proposed for hydrocarbon emission control.

4.9   **OZONE**

Ozone is a clear, blue gas with a sharp odor. Trace amounts occur naturally in the upper atmosphere. The maximum concentration is found at an approximate height of 25 kilometers (15.5 miles). Ozone molecules are made of oxygen atoms.

Single oxygen atoms are too reactive to exist alone for any length of time. They readily combine with other oxygen atoms or with a multitude of other atoms or molecules. Life-sustaining oxygen is composed of diatomic molecules—each formed from two oxygen atoms bonded together. Ozone is composed of three oxygen atoms bonded together. It is sometimes produced by lightning passing through atmospheric oxygen. However, most of the atmospheric ozone forms at high altitudes from the action of solar ultraviolet energy on oxygen. This region of the atmosphere, called the *ozone layer*, provides vital protection for living organisms on the earth's surface. It absorbs nearly all of the ultraviolet energy from the sun. You have probably experienced the burning effects of the small portion of this energy which does reach the earth's surface. If you were to stare directly at the sun, this same energy could blind you. Fortunately, the ozone acts as a screen, filtering out potentially dangerous quantities of ultraviolet radiation. Obviously, the ozone naturally produced in the upper atmosphere is highly beneficial. By emitting nitrogen dioxide into the air, however, man has caused the production of abnormal amounts of ozone in the lower atmosphere by photochemical reaction.

Effects. Ozone is a powerful oxidant and readily attacks other substances. It is highly poisonous to living organisms. At low concentrations, it produces chest pain, coughing, and often eye irritation. Prolonged exposure apparently increases susceptibility to bacterial infections. Although small dosages of ozone were previously considered healthful, research has shown that as little as 1 ppm is highly irritating after continued exposure. Research rodents, which endured 1 ppm ozone for eight hours daily during a single year, developed ailments such as bronchitis, fibrosis, and bronchiolitis. Continuous exposure to ozone considerably shortened the lives of laboratory guinea pigs. Concentrated ozone can kill both plants and animals.

Ozone attacks the upper surface of the leaves of green plants, causing spotting. Many crops such as grapes, sugar beets, spinach, oats, lettuce, alfalfa, and tobacco are highly susceptible. The variable sensitivity of different plants is apparently related to the rate of metabolic and photosynthetic functions of the plants during their exposure to ozone.

Ozone attacks textiles; clothing gradually discolors and disintegrates. It also has a great effect on rubber, automobile tires in particular. A simple test for ozone content in the atmosphere is to observe the cracking effect on a piece of stretched rubber tubing. The rate of rubber deterioration is a measure of the ozone concentration. This latter effect is particularly important to aircraft which fly at high altitudes where the ozone concentration is usually high. Ozone can damage not only the rubber tires on the aircraft wheels but also the rubber sealing around the windows and the rubber insulation on the plane's electrical wiring.

The problem of ozone is yet another example of both unexpected and undesirable consequences resulting from the indiscriminate contamination of our atmosphere. Before such damaging concentrations of ozone can be eliminated, emissions of other chemical reactants, in particular nitrogen dioxide, must be controlled.

For Thought and Research

1 (a) Would you expect significant quantities of ozone to be produced in your area? If so, is ozone damage a recognized problem there? Explain your answer in terms of the factors necessary for ozone production in the lower atmosphere.

  (b) Contact your local air pollution authority to verify your prediction.

2 Ozone can be easily prepared using the following method: Set up an induction coil with a spark jumping a gap between two points. The characteristic odor in the vicinity of the discharge is produced by the ozone formed. This same process occurs during an electrical storm.

3 You can investigate the effect of ozone on various materials by following the procedure outlined in Section 7.9.

Recommended Readings

1 "Air," *The Pollution Reader* by A. DeVos et al., Harvest House, 1968.
2 "The Control of Air Pollution" by A. J. Haagen-Smit, *Scientific American,* January, 1964. This article relates ozone concentrations to photochemical smog studies in Los Angeles.
3 *The Chemistry of Urban Atmospheres* by L. G. Wayne, Technical Progress Report Vol. III, A.P.C.D., Los Angeles County, December, 1962. This report also investigates the role of ozone in photochemical smog production.

4.10 LEAD

Lead aerosol is a familiar contaminant in urban atmospheres. Extensive controversy prevails over the health effects of the current concentrations. In this section, consider the possible consequences to the entire ecosystem.

(a) Sources. Lead is a natural component of air, water, and soil, as well as of plant and animal life. Naturally-occurring levels are not well defined. Silicate dusts from soil erosion and volcanic emissions are the chief natural contributors of lead to the atmosphere. However, man accounts for the greatest emission of lead into the air. Sources are coal combustion, manufacturing, pesticide spraying, waste incineration, and, most important in urban areas, the use of leaded gasolines.

The compression ratio of an automobile engine determines to a large extent the octane rating of the gasoline required to produce maximum engine power and performance. In order to meet the high octane requirements and to prevent engine "knock" or preignition, tetraethyl lead and tetramethyl lead compounds have been added to gasoline products. The only alternative is a costly conversion by oil companies to more extensive processing of petroleum.

Almost two-thirds of the lead consumed in gasoline is exhausted to the atmosphere. Between 25 and 50% of these lead particles remain airborne. The number of cars is increasing. Also, the trend has been toward engines requiring gasoline with a higher octane rating. Hence the use of gasoline with a higher lead content has become widespread. This has increased airborne lead concentrations to potentially dangerous levels in many cities. Engines of lower compression, which require lower octane gasoline, are currently being designed. These new engines will operate effectively on lead-free gasoline—the product of the future. Yet, many are convinced that this conversion will prove to be both costly and unnecessary. Some argue that such low compression engines may require more gasoline per mile and natural petroleum resources could be depleted at a 6% higher rate than at present. In any case, a drastic reduction in the number of gasoline-powered automobiles is the most valid remedy to the entire problem.

Core samples from the Greenland ice cap reveal an increase in lead content of 400% between 1750 and 1940, and a further increase of 300% between 1940 and 1967. However, the content of sea salts in the same ice cap has not changed during this same time period. This indicates that natural fluctuations in the quantity of materials deposited in the ice are not responsible for the accumulation of lead. Furthermore, the lead content has not increased in a similar pattern in the ice of the Antarctic. One strongly supported theory maintains that atmospheric lead pollution is causing this deposition. Increased atmospheric lead content first began with lead smelting and was further stimulated by the expanding use of leaded gasolines. Both of these contamination sources are concentrated in the northern hemisphere. The analysis of sea water, sediments, and even snow from remote areas indicates that the atmosphere of the northern hemisphere now contains 1,000 times more lead than would naturally occur without man's contributions. Since the global circulation of air masses does not carry this lead pollutant into the southern hemisphere to Antarctica, the ice there does not contain the same deposits as the Greenland ice cap.

(b) Effects. How does this toxic element affect living organisms? Even if the present levels of lead are not directly harmful to human health, the concentration of lead steadily increases in food chains. This could ultimately produce a toxic dose for man and other organisms which feed on large quantities of food contaminated with lead. Lead is accumulating in plant life, particularly near highways or where pesticides are widely used.

Most of the lead to which people in urban areas are exposed is airborne. In 1924, less than one million pounds of tetraethyl lead were consumed yearly in automobile fuel. By 1968, 700 million pounds of tetraethyl lead were consumed in automobile fuel. Aviation gasoline used another 50 to 60 million pounds. Those who oppose the removal of this compound from gasolines should note that the average North American now carries in his system 100 times more lead than was common before the introduction of tetraethyl lead.

Airborne lead enters many phases of man's environment. The level of lead in the rainfall of two major American cities is twice as high as that allowed in drinking water. Furthermore, a high correlation exists between the concentration of lead in the rainfall and the sales of leaded gasoline. These rainfalls are carrying lead into the North Pacific and North Atlantic 50 times faster than nature introduced it in the past. This lead fallout, which amounts to 500,000 tons yearly, could eventually upset the mineral nutrient balance of the oceans and produce lead poisoning. Airborne lead has also been blamed for unusual rain production. If lead combines with iodine high in the atmosphere, the particles can, under the proper conditions, produce the same results as the silver iodide crystals often used for cloud seeding.

Before modern safety standards were established, lead poisoning was a common occupational disease. Industrial workers inhaled dusts or vapors of inorganic lead compounds. Painters exposed to lead-base paints often absorbed poisonous amounts of lead compounds through their skin. In 1968, the average lead concentration in the bloodstream of North Americans was approximately 0.25 ppm. This figure is just less than half of the level at which people who work with lead in industry are removed from further exposure. Traffic policemen, garage mechanics, and parking lot attendants have an average of 0.36 ppm of lead in their bloodstreams. In 1965, authorities warned that North Americans were in danger of severe chronic lead poisoning. And lead pollution is expected to increase in the future.

Lead and its compounds affect the central nervous system. Symptoms of chronic lead poisoning include headaches, loss of appetite, dizziness, insomnia, anemia, weakness, and miscarriage. The greatest danger is the harmful effect of lead on the red blood cells. Lead changes their size and shape, and makes them brittle. Lead dustfall in some residential areas has been correlated to a high incidence of heart disease. Acute lead poisoning produces stupor, then coma, and, eventually death. Lead is a cumulative poison. Although chronic lead poisoning is very difficult to diagnose, many authorities fear that low-level effects already exist. How much longer can we afford to wait before lead emissions are strictly controlled?

For Thought and Research

1 (a) Is lead-free gasoline available at service stations in your area? If so, is this gasoline more expensive than leaded gasoline?

(b) Consult the local service station attendants to determine the percentage of motorists who are now using lead-free gasolines.

2 Are lead-containing pesticides used in your area? Is lead accumulating near the major thoroughfares or highways in your area? Are the local service station attendants exposed to higher levels of lead? You can investigate by analyzing dustfall for lead content as outlined in Section 7.5.

3 The installation of pollution emission controls and the conversion to engines of lower compression will lower the power of future cars. Are high-powered cars really a necessity or do they simply represent a life-style which is helping to destroy the environment?

Recommended Readings

1 *Cleaning Our Environment. The Chemical Basis for Action*, American Chemical Society, 1969. See the section dealing with lead for a more detailed analysis of lead additives in gasoline.

2 "The Air We Breathe," *Pollution Probe* by D. A. Chant, New Press, 1970.

3 "Solids in the Air," *This Vital Air, This Vital Water* by T. G. Aylesworth, Rand McNally, 1968.

4 "Our Ecological Crisis," *National Geographic*, December, 1970.

4.11 OTHER POLLUTANTS

Fluorides. Highly toxic fluoride compounds, whether discharged into the atmosphere in gaseous or particulate form, present a very serious problem. The major sources of fluorides are industries which manufacture aluminum, steel, and phosphate fertilizers. Brick plants, glass, pottery, and ferro-enamel works also emit fluorides. Smaller amounts are released during coal combustion. Even at concentrations as low as 0.001–0.10 ppm, hydrogen fluoride (HF) and other volatile fluorides can produce severe effects on the environment.

Plant leaves have tiny surface pores called stomata. These open during the day to permit the exchange of gases between the inner cells of the leaves and the atmosphere. Gaseous hydrogen fluoride accumulates in plant leaves. When animals feed upon these plants they acquire a disease called fluorosis. Cattle grazing on fluoride-contaminated grass develop uneven teeth which make chewing difficult. Also, their joints become so swollen that many cannot tolerate standing.

Many plants, such as peach, apricot, prune, and gladiolus are extremely sensitive to hydrogen fluoride, even at concentrations as low as 0.02–0.05 ppm. In semi-rural Florida, phosphate plants spew as much as 7 tons of fluorides a day on neighboring citrus groves, ranches, and gladiolus farms. When orange and lemon trees absorb these fluorides, they produce smaller yields; gladioli turn brown and die. Crops that ordinarily mature in 80 days now take up to 6 months. Across the nation, fluoride emissions have severely affected agriculture.

Gaseous fluoride compounds can etch transparent glass surfaces until they resemble frosted glass. They also attack paint and metal surfaces. They can also increase the fluoride consumption of people who already drink fluoridated water. In addition, foods and beverages processed with fluoridated water have increased fluoride concentrations which are passed on to the consumer. The "officially approved" level of fluoride consumption is low but even this "safe" concentration is now considered by researchers to be dangerous for certain people. Moreover, this level is now exceeded in many communities affected by fluoride pollution. Both fluoride pollution and fluoridation of water must be monitored more carefully.

Asbestos is a hydrated magnesium silicate which has almost universal use in industry—all the way from the roofing, through the ceiling, down to the floor tiles. World demand for asbestos has increased from the 1934 consumption of 300,000 tons to the more than 2,400,000 tons used in 1961. Asbestos is found in pipe and electric insulations, brake linings, and, unhappily, the human lungs.

Recent studies reveal that inhaled asbestos fibers often become embedded in the lung tissue where they apparently cause mechanical irritation. This may result in a rare tumor form called mesothelioma and also in lung cancer. Although many asbestos handlers have developed mesothelioma, most victims are persons living close to, but not working in, the asbestos industry. Authorities hope that autopsy studies from the asbestos-mining regions of Quebec will help determine the importance of exposure to asbestos.

When you consider the proportion of the population which is exposed daily to the emissions from industry you can better appreciate the value of this constant research. It is by far a simpler task to insure the purity of the food and water which we consume than to guarantee the quality of the atmosphere which we breathe. Yet air is our most critical commodity.

For Thought and Research

1 Are fluoride emissions a problem in your community? If so, what are the major sources of this pollution? What effect has this contamination had on the area? What action has been taken to control the problem?

2 Is the drinking water in your community fluoridated? Is this water used to process the food and beverages which you consume?

3 Should farmers who suffer damages due to fluoride contamination be recompensed by the source responsible for the emissions? Would the necessary legal action prove difficult? Why? If you were the community leader, how could you protect the rights of industry and citizens alike?

4 Is there a great deal of industrial activity in your neighborhood? Are you exposed to airborne asbestos from building sites or any other source? You can trap and identify airborne asbestos using the methods outlined in Sections 7.2 and 7.3.

Recommended Readings

1 "Air," *The Pollution Reader* by A. deVos et al., Harvest House, 1968.
2 *Air and Water Pollution*, by G. Leinwand, Washington Square Press, 1969. See the selection "No Place to Hide" dealing with fluoride contamination.
3 "Menace in the Skies," *Time*, January 27, 1967.

4.12 NATURE AT WORK

The concentration of air pollutants in any region depends greatly upon the characteristic weather conditions. Wind speed and direction and the vertical temperature distribution are the principal elements. Moisture also plays a role.

The Effects of Wind. The rate at which air pollutants are carried away from their sources depends upon the wind speed at or near the point of release. Concentrations of contaminants are diluted and dispersed both horizontally and vertically into increasingly larger volumes of air by the turbulence of the wind. The turbulence or gustiness of a moving air mass increases with its average speed. Thus, high winds most rapidly reduce the concentration of pollutants while moving them far from their origin. When low wind speeds prevail, contaminants tend to remain near their source. Their concentrations increase steadily throughout periods of very light winds or calm. Little mixing with the surrounding air takes place. The high concentrations of pollutants simply drift slowly into neighboring areas.

Surface features of the land can often influence local wind speed and direction. For instance, winds moving across a mountain valley are often channeled in the direction of the valley. Light winds funneled into a pass increase in speed for some distance. Such factors greatly influence both the rate of pollutant dispersion and the direction in which it occurs.

Temperature Inversions—A Warning from Nature. Seasons of low temperature bring increased fuel consumption for the heating of homes and businesses. This leads to greater emission of pollutants. While temperature alone does not affect the eventual dispersal of contaminants, the change in temperature with altitude is a controlling factor, for this determines the stability of the atmosphere. Ordinarily temperature decreases rapidly with altitude, and the air mass is unstable. Vertical turbulence carries pollutants high into the cooler air above. However, under certain conditions, the temperature shows little or no decrease with height. It may actually *increase* with altitude. Such an air mass is stable. Vertical movement is restricted. What happens, then, to the concentrations of pollutants produced at ground level?

On October 26, 1948, citizens of Donora, Pennsylvania, awoke to a dark, ominous morning. The various industries and businesses of the town were emitting their usual contaminants into the air. On any ordinary day, warm, light air rising from the ground surface carried such pollutants into the cooler atmosphere above. Here they were diluted and dispersed by prevailing winds. However, on this particular day, a blanket of warm air had formed above the cooler air mass enveloping the town. During the day, warm surface air and vapors produced by industries steadily cooled while rising. When this air reached the upper limits of the cold air mass, it was no longer lighter than the warmer air mass above. Unable to rise further, the contaminated air was trapped between the upper and the lower air masses. The cold surface air mass captured and suspended the pollutants. Gradual cooling of the stagnant air mass resulted in a steady deposition of filth on the town. Within 48 hours, visibility was drastically reduced. Donora medical authorities were swamped with people suffering from a range of ailments: stinging eyes, throat irritation, nausea, breathing difficulties, nasal congestion, and coughing. During the four long days that the deadly shroud hung over Donora, almost 50% of the 12,300 citizens suffered illness. Before a heavy rain eventually cleared the air, death claimed 20 people and numerous animals such as cats, dogs, and birds.

Meteorologists define the abnormal atmospheric conditions which triggered this episode as a *temperature inversion* (Fig. 4-13). Instead of the normal decrease in air temperature with elevation, an exceptionally stable layer of cold air is trapped under a layer of warmer, less dense air. The air temperature increases with altitude. The normal air circulation and turbulent mixing which produce winds are suppressed. As a result, concentrations of contaminants steadily accumulate in the stagnant air until the weather changes. Geographical factors, seasonal temperature changes, and wind direction and velocity influence the frequency of such inversions in any area.

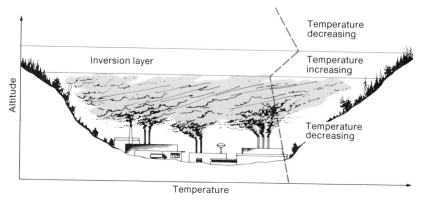

Any conditions which make the air at or near the surface cooler than the atmosphere immediately above produce a stable air mass. On a clear night, the heat absorbed by a land surface during the day is lost by radiation to the upper atmosphere. The surface air mass then becomes cooler than the atmosphere above. Thus, stable conditions prevail during the early morning hours in clear, calm weather. After sunrise, the surface air is rapidly heated. The temperature inversion which developed during the night dissipates. Unless this stable period is extended, air pollutants rarely persist for more than a few hours.

Water has a much greater capacity than land for retaining heat. Therefore a body of water tends to moderate the temperature of a surrounding air mass. Bordering land areas have less temperature variation between day and night. Consequently, they are less subject to early morning stability conditions. If, however, a warm air mass over a body of water moves inland and meets a colder air mass over the land surface, the warmer, lighter air mass is forced above the cooler air. A temperature inversion is formed. Such conditions occur most frequently over valleys and near coastal regions, but any area can be affected.

When surface heat radiates from the slopes of a valley under a clear night sky, the denser, cooler air next to the ground flows down into the valley. If the bottom end of the valley opens into lower levels, the cold air drains out as long as surface cooling continues. During the day, when the valley sides are warmed by the sun, the process is reversed. Why?

If a valley is totally enclosed to form a basin, cold air draining down to the lowest level forces warmer air upward (Fig. 4-14). Throughout the night, an increasing layer of cold air settles in the basin. Sufficient heat from the sun during the following day should gradually disperse this cold air mass. However, fog often forms in such a basin of cold air. This surface "cloud" layer reflects the sun's rays. Hence a cold, damp, and possibly

Fig. 4-14
Cold air masses flow down the sides of a valley, forcing the warmer air mass at the bottom upwards.

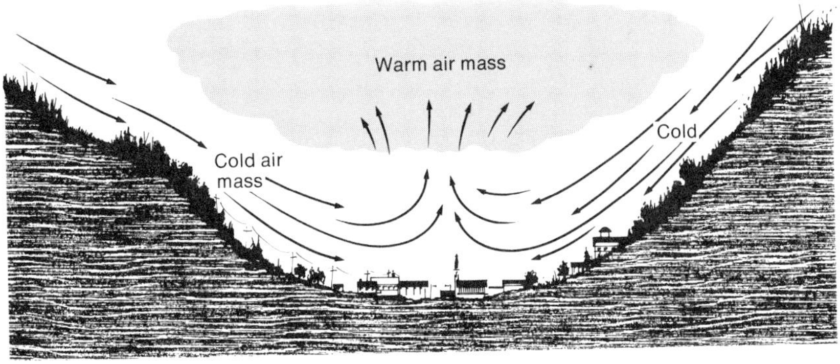

Warm air mass

Cold air mass

Cold

polluted air mass persists until brisk winds arrive to clear the fog. However, if a warm air mass should flow over the hills and across the top of the valley, the cold basin air becomes trapped in a temperature inversion. This situation occurs most often in the late summer and autumn. Why?

Wind velocities associated with temperature inversions are generally 7 mph or less. In Niagara County, New York, the wind velocity is usually greater than 8 mph, with a yearly average of 11.6 mph. Thus any temperature inversions which occur there are short-lived. Such inversions are recorded during 20 to 25% of the total hourly observations in Niagara County. Inversions have a 30% occurrence in Albany and Syracuse, New York, and a 40% occurrence along the Pacific Coast. Suggest possible reasons for these variations.

Temperature inversions present a genuine threat to increasingly industrialized cities with problems of air pollution. The Donora tragedy is only one example. The worst recorded air pollution disaster took place in London, England, during December, 1952. A temperature inversion over the city produced a five-day nightmare. Steady contributions of soot and smoke from the coal-burning city reduced visibility to less than a yard within three days. Before a passing cold air mass cleared the terrible smog, 4,000 people had died as a direct result of pollution. During the following two months, another 8,000 people died from illnesses directly attributed to this killer smog. Since 1953, New York City has suffered four critical pollution episodes during temperature inversions. Alarmed, cities throughout North America have instituted devices to monitor air pollution levels. Various warning systems have been established. In some areas, such as Los Angeles, conditions often reach the point where children are advised not to run and play in the contaminated air, lest they breathe too often and too deeply. Is this the answer to our problems?

The Effects of Moisture. Moisture, including rain, snow, or sleet, does not affect pollutant concentrations as directly as wind or atmospheric stability. Although precipitation is generally regarded as an air cleansing agent, it is very inefficient. Large raindrops, which are most effective, normally accompany the strong winds and atmospheric turbulence which minimize air pollution problems regardless of moisture. However, light rain or drizzle during a period of light wind has little cleansing effect on the atmosphere.

The moisture content of the air is an important factor in the deterioration of materials caused by various pollutants. In addition, fog usually limits the amount of solar heat energy reaching ground level. This, in turn, affects the vertical dispersion of airborne contaminants. Moreover, air contaminants may directly affect the weather by providing condensation nuclei for fog formation or abnormal production of rain. Increasing air pollution could conceivably affect global weather patterns. Since climate is a controlling factor for all life on earth, control of air pollution may well be man's greatest challenge.

For Thought and Research

Regardless of its present state of development, consider the area in which you live as a possible industrialized community. Consult the local weather bureau for any data which you may need to answer the following questions.

1 Do the speed and direction of the prevailing winds favor a rapid dispersal of airborne particles in your area?

2 Does a seasonal variation exist in the prevailing wind speed and direction? If so, how does this alter the dispersal of air pollutants?

3 Do local surface features alter wind speed and direction in your area?

4 Is your community affected by pollutants produced by some neighboring region? Why?

5 Would airborne contaminants produced in your area greatly affect some neighboring region? Why?

6 Does the temperature fluctuate greatly between average daily and nightly readings in your area? Are you located near a temperature moderating body of water?

7 Do local surface features favor the formation of temperature inversions? Explain your answer fully.

8 Are temperature inversions common in your area at any time of the year? Why?

9 Would occurring temperature inversions persist long in your area? Could they create a serious air pollution episode in your community as it actually exists?

10 Would moisture factors in your area play an important role by either washing pollutants from the air or trapping them by fog formation? How would excess air contamination affect the weather patterns in your region?

---

1 "The Deadly Sky," *This Vital Air, This Vital Water* by T. G. Aylesworth, Rand McNally, 1968.
2 "Air Pollution and Population Growth," *Population, Resources, Environment* by P. R. Ehrlich and A. H. Ehrlich, W. H. Freeman, 1970.
3 "Menace in the Skies," *Time*, January 27, 1967.
4 "Our Ecological Crisis," *National Geographic*, December, 1970.
5 "The Nature of Photochemical Smog," *The Chemistry of Urban Atmospheres* by L. G. Wayne, Technical Progress Report Vol. III, A.P.C.D., Los Angeles County, December, 1962.
6 "Meteorology and Topography," *Air Pollution—Niagara County*, State of New York Air Pollution Control Board, Albany, N.Y., 1964.
7 "Air Pollution" by N. Hinch, *Journal of Chemical Education*, February, 1969.

4.13   PHOTOCHEMICAL SMOG

Photochemical smog refers to a specific type of polluted atmosphere most commonly associated with the Los Angeles area. Los Angeles had almost no pollution problem until an explosive growth in population and industry began around 1940. Suddenly the attractive, clear air was transformed into a brown haze laden with smoke and offensive odors. The city lies in a mountain-rimmed "bowl" which traps pollutants during frequent temperature inversions. Studies reveal that the renowned California sunshine is helping to make life in Los Angeles unpleasant and, more important, hazardous to health. Under certain conditions, sunlight promotes chemical reactions between airborne pollutants to produce a damaging and irritating haze. Man has, in effect, produced a "factory in the sky" (Fig. 4-15, page 140).

Raw Materials. Pollutants produced directly from natural or urban sources are called primary contaminants. Any pair of substances which react when mixed in a laboratory should, theoretically, react when mixed in the atmosphere. Fortunately, most of the common pollutants are generally inert toward air. Also, they are usually sufficiently diluted by the atmosphere to prevent rapid reaction with other pollutants—unless activated by an energy source such as the sun. Some primary contaminants, however, react readily with each other or with normal atmospheric components such as water vapor. These primary contaminants, the reactive hydrocarbons in particular, provide the raw materials for photochemical smog production. They are primarily the result of gasoline vapors and automobile exhaust.

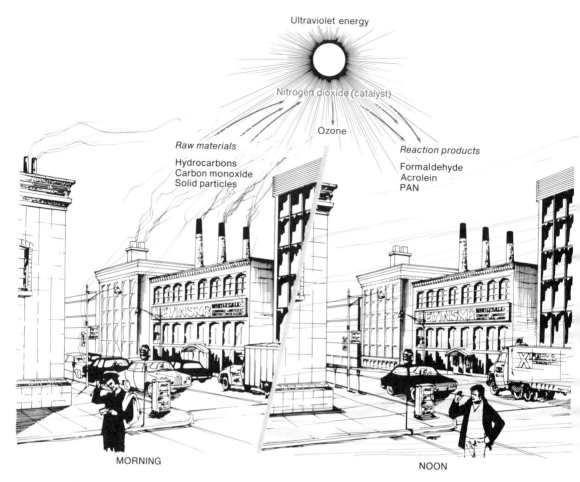

Ultraviolet energy

Nitrogen dioxide (catalyst)

Ozone

Raw materials

Hydrocarbons
Carbon monoxide
Solid particles

Reaction products

Formaldehyde
Acrolein
PAN

MORNING

NOON

Reaction Catalyst. Nitrogen dioxide acts as a catalyst in the production of photochemical smog. A catalyst is a substance which speeds up the rate of a chemical reaction without being used up. Nitrogen dioxide absorbs the ultraviolet portion of sunlight and converts it into chemical energy which, in turn, promotes reactions between the hydrocarbons. Each nitrogen dioxide molecule uses the ultraviolet energy to change to a molecule of nitric oxide. As it does so, it gives off an energized (highly energetic) oxygen atom.

Fig. 4-15
"Factory in the Sky"— photochemical smog production.

Nitrogen Dioxide $\xrightarrow{\text{ultraviolet energy}}$

Nitric Oxide + Energized Oxygen Atom

$NO_2 \xrightarrow{\text{ultraviolet energy}} NO + O^*$

Later in the reaction, the nitric oxide molecules each regain an atom of oxygen. Hence the original supply of nitrogen dioxide is not diminished.

140   Air Pollution

Reaction Products. The energized oxygen atoms quickly react, either with oxygen molecules to produce ozone, or directly with the hydrocarbons to form secondary contaminants. The ozone can also react with the hydrocarbons to produce secondary contaminants. The secondary contaminants are responsible for the damaging effects of photochemical smog. Although a complete identification of these products has not yet been made, eye irritants such as formaldehyde, acrolein, and PAN (peroxyacylnitrate) have been recognized. Other aldehydes and ozone are also present in significant concentrations.

Reaction Conditions. Photochemical smog usually occurs on dry, windless days which are warmer than usual. The rate of development and the resultant intensity of the photochemical smog depend upon certain reaction conditions:

(a) The concentration of pollutants and their relative proportions. This is determined by meteorological conditions as well as by the volume of traffic and the level of industrial activity.

(b) The vertical temperature distribution. Temperature inversions, common to Los Angeles, prevent the dispersion of primary contaminants into the upper atmosphere. The accumulated hydrocarbons and nitrogen dioxide increase the probability of smog production.

(c) Air turbulence. Low wind speeds, associated with temperature inversions, limit dispersion of pollutants. During the warm months in Los Angeles, a transition period of offshore to onshore air flow occurs shortly after sunrise. This produces a stagnant air mass just as the morning traffic rush begins.

(d) Intensity and duration of sunlight. When suitable meteorological conditions prevail, the concentrations of primary contaminants in Los Angeles are sufficient to produce an objectionable smog within a 1 or 2 hour exposure to bright sunlight.

Effects of Photochemical Smog. As traffic increases during the early morning, the clear city air gradually becomes a yellow-brown haze with a characteristic odor. The shroud of pollution obscures prominent landmarks and limits visibility. The smog intensity increases and its properties change during the day, depending upon the movements of the polluted air mass. A peak intensity usually prevails in downtown Los Angeles around noon. During severe episodes most people suffer eye irritation and many are also affected in the nose and throat region. If such photochemical smog occurs on several successive days, the effects become more widespread and serious. Often, the arrival of a sea breeze or an end to the temperature inversion produces a sudden clearing of the smog.

The air above Los Angeles County receives almost 15,000 tons of contaminants daily. The inhabitants no longer dwell in a sea of air—they live in a sewer. Each person daily inhales 6,000 gallons of this mixture which contains more than 50 different contaminants, many of which are carcinogens. Along with the secondary contaminants, arsenic, asphalt, benzol, carbon black, chromium, creosote oil, mustard gas, paraffin oil, and tar have been identified as smog constituents.

What happens to the people who constantly breathe this poisoned air? In 1962, a state report revealed that more than 2,000,000 Californians—1 person in 8—suffered to some extent from a chronic respiratory problem such as emphysema, bronchitis, or asthma. Of these, 670,000 (400,000 of whom worked in Los Angeles County) were disabled to some degree as a result; 25,000 were permanently and totally disabled. Furthermore, 3,000 of these were expected to die during the year. The death rate from emphysema increased four-fold in the 10 years preceding this 1962 report. The trend continues today.

"Emphysema" is a Greek word meaning "swelling or inflation due to the presence of air." The lungs are composed of more than 600 million tiny air sacs called alveoli. These cup-shaped hollows are surrounded by thick networks of blood vessels. The thin alveoli walls permit easy passage of gas molecules to and from the bloodstream. The alveoli provide a total surface area within the lungs of an estimated 1,000 square feet—the size of a tennis court! Emphysema causes an abnormal enlargement of the tiny alveoli which eventually breaks down their thin walls. As a result, the effective surface area of the lungs is gradually reduced to the size of a Ping-Pong table. As the entire lung system balloons up, the victim ends up with a flabby, ineffective "bellows system." A patient with severe emphysema does not have enough breath to blow out a match! Furthermore, the damage is irreversible. Doctors rarely encounter an emphysema patient who is not a smoker. Because of the high concentration of irritants directly inhaled, cigarette smoking is air pollution at its worst. The resultant lung damage makes smokers highly vulnerable to the effects of smog.

There is more lung disease in southern California than in any other state with the exception, perhaps, of Arizona or Florida where patients are sent for relief from the symptoms of their diseases. Despite stringent regulations in Los Angeles designed to curtail pollutant emissions, the growing population, having rejected a rapid transit system, relies heavily on the leading culprit—the automobile. Further, the predicted automobile increase in Los Angeles is expected to more than nullify any advances in emission controls for car engines of the future. As the smog persists, doctors continue their warnings. In one year alone, more than 10,000 patients were advised to leave the region. A survey revealed that a third of the doctors were tempted to do the same.

Smog damage to vegetation has been a growing problem since 1942. Various plant species react differently with respect to the type and degree of injury. Although the full extent of the damage is often not immediately evident, many crops are eventually rendered unmarketable. The productivity of citrus groves and truck farms in the Los Angeles area has greatly declined. Some plants, such as spinach and orchids, are impossible to grow in metropolitan Los Angeles. Extended periods of severe smog have produced serious alfalfa crop failures.

Although Los Angeles remains the chief center for photochemical smog, this phenomenon has been detected in varying degrees of intensity in most major metropolitan areas in recent years. Extensive research has revealed a great deal concerning its cause and mechanism. Unless pollutant emissions are controlled, the sun, which has long made life on this planet possible, could easily make life in a city an atmospheric nightmare.

For Thought and Research

1 Evaluate your area as a potential site for photochemical smog formation by examining each of the following factors:

(a) *Raw Materials.* What proportion of the population relies for transportation on private automobiles rather than public transport? Do air pollutants present a measurable problem in your area? If so, which pollutants in particular?

(b) *Reaction Catalyst.* Are significant concentrations of nitrogen dioxide present?

(c) *Reaction Conditions.* Are temperature inversions common? What is the average wind velocity? Are calm periods frequent? How does the intensity and duration of sunlight compare with areas such as Los Angeles which are afflicted by photochemical smog? Is photochemical smog formation more probable in one season than another?

2 If you were attempting to solve the problem of photochemical smog in the Los Angeles area, what changes or legislation would you introduce? If photochemical smog is not yet a problem in your city or community, what safeguards should be taken for the future?

Recommended Readings

1 *The Chemistry of Urban Atmospheres* by L. G. Wayne, Technical Progress Report Vol. III, A.P.C.D., Los Angeles County, December, 1962.
2 "Photochemical Smog: Cause, Effects and Cures," *Profile of Air Pollution Control in Los Angeles County*, A.P.C.D., Los Angeles County, 1969.

3 "Menace in the Skies," *Time*, January 27, 1967.

4 "Air Pollution and Population Growth," *Population, Resources, Environment* by P. R. Ehrlich and A. H. Ehrlich, W. H. Freeman, 1970.

5 *Cleaning Our Environment. The Chemical Basis for Action*, American Chemical Society, 1969. Read the section dealing with hydrocarbons.

6 "The Control of Air Pollution" by A. J. Haagen-Smit, *Scientific American*, January, 1964. An excellent account of photochemical smog.

7 *The Poisoned Air Around Us* by V. Boesen, U.S. Department of Health, Education, and Welfare, 1967. Distributed by the National Center for Air Pollution Control, Bureau of Disease Prevention and Environmental Control. An account of the health effects produced by smog.

8 "Gases in the Air," *This Vital Air, This Vital Water* by T. G. Aylesworth, Rand McNally, 1968.

9 "Our Ecological Crisis," *National Geographic*, December, 1970.

10 "Human Energy Production as a Process in the Biosphere," *Scientific American*, September, 1970.

11 "Air," *The Pollution Reader* by A. deVos et al., Harvest House, 1968.

4.14 THE FIRST LINE OF DEFENSE— AIR QUALITY STANDARDS

About 700 years ago—in 1273—King Edward I of England passed a decree restricting the use of fire. This was the first recorded attempt to control air pollution. Later, in 1306, the English government passed another law forbidding anyone to burn coal while Parliament was in session. According to legend, one defiant soul was actually hung for breaking this law. If present legislation could be as stringently enforced, the problem of pollution would soon diminish—either through strict observance of the law or a considerable reduction of the general populace!

Air pollution is not a new problem but simply a rapidly increasing one. The city of Tokyo, Japan, has four times the incidence of bronchitis per thousand citizens as the rest of Japan. School children in Yokkaichi, Japan, must often spend the day in a surgical mask impregnated with chemicals which protect them from the overwhelming industrial pollution. American troops stationed in parts of the Japanese mainland have developed respiratory diseases associated with the severe air pollution. From countries around the world—Greece, Spain, France, Germany, Italy, Britain, Brazil, Chile—come similar tales of damaging coexistence with air pollutants. How long can this situation be tolerated? Is there, in fact, any overall solution?

During the past decade, the general public has finally become aware that smog is far more than a necessary evil which, like increasing income tax, must be accepted in return for technological progress and, presumably, a better quality of life. But now we cannot "see the forest *or* the trees." Apart from the property damage, which defies estimation, is the irreparable injury to man and other living organisms. Air pollution is, at best, a health hazard—at worst, a killer. What is being done?

Los Angeles is commonly known as the smog capital of the U.S.A. It is also a pioneer in the fight for cleaner air. As early as 1948, Los Angeles County began limiting the dirt and fumes released by steel factories, refineries, and countless smaller industries. The use of more than a million home incinerators was banned; burning in public dumps was forbidden. The result? Monthly dustfall, which in some areas had amounted to 100 tons per square mile, was reduced by 66% to the level which existed in 1940 (before increasing population and industrial growth created severe smog problems). Intensive research has gradually disclosed the complex mechanism of photochemical smog production. Corresponding steps have been taken to reduce the available reactants. Control measures are now directed at the release of hydrocarbons, carbon monoxide, and nitrogen oxides. For example, a 1951 Los Angeles district survey showed that refinery losses amounted to more than 400 tons of contaminants *daily*. Subsequent control measures have lowered this figure to an estimated 85 tons—still a tremendous discharge, but an improvement.

However, attempts at pollutant reductions have been offset by an increase in the emissions from vehicular traffic (Fig. 4-16). When the automobile emerged as the major source of air

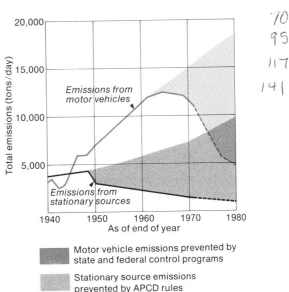

Fig. 4-16
The relationship between motor vehicle emissions and stationary source emissions in Los Angeles County, 1940–1980. (Data published by Air Pollution Control District, Los Angeles County.)

TABLE 13

Level	Effects	Pollutant concentration (ppm)	
"Adverse"	Sensory irritation and damage to plant life	Sulfur dioxide	1
		Oxidant index	0.15
		Hydrogen sulfide	0.1
"Serious"	Danger of altered bodily function or of chronic disease	Carbon monoxide	120
		Sulfur dioxide	5
		Hydrogen sulfide	5
"Emergency"	Acute sickness or death is possible among sensitive persons	Carbon monoxide	240
		Sulfur dioxide	10

pollution, the California Department of Public Health adopted the first air quality standards as outlined in Table 13.

These standards provided a basis for pollution control legislation, aimed in particular at the automobile industry. Beyond these efforts, other alternatives to gasoline combustion must be considered. For example, monitored carbon monoxide readings in downtown Los Angeles show distinct peaks resulting from rush-hour traffic. Why, then, are the number and use of cars not restricted?

The establishment of air pollution controls in any community is complex and difficult. An intensive program must be developed and close cooperation between government, industry, and other agencies is essential. For example, industries will be understandably reluctant to install smoke curtailing devices as long as cities continue to burn trash in open dumps. Cost is also a critical factor. For example, the Detroit city council would not ban the burning of leaves because the city presumably could not afford to haul the leaves off to dumps. But who pays the cost, both discernible and hidden, of the subsequent air pollution?

Many areas have developed air pollution indices which are a measure of different air contaminant concentrations. Formulating an index that presents a valid indication of the total air contamination is difficult. Obviously not all pollutants can be monitored simultaneously. Different cities tend to monitor the most prominent and critical pollutants in their own area. Table 14 indicates the pollutants monitored by a few North American cities.

In the province of Ontario, levels of sulfur dioxide and particulate matter are measured daily. The correlation between the level of particulate matter, commonly referred to as the

TABLE 14 AIR POLLUTANTS MONITORED FOR AIR POLLUTION INDICES (■)

City	Solid particles	Sulfur dioxide	Carbon monoxide	Nitrogen oxides	Oxidants
Buffalo, N.Y.	■				
Detroit, Mich.	■				
Hamilton, Ont.	■	■			
Los Angeles, Cal.		■	■	■	■
New York, N.Y.	■	■	■		
San Francisco, Cal.	■		■	■	■
Sarnia, Ont.	■	■			
Toronto, Ont.	■	■			

coefficient of haze (COH) and the level of sulfur dioxide is poor. Thus, the pollution index value must be a mathematical combination of both sulfur dioxide (SO_2) and particulate matter concentrations. For example, the Sarnia, Ontario, index is calculated as follows:

$$\text{Air Pollution Index (A.P.I.)} = \frac{SO_2 \text{ Index } + \text{ COH Index}}{2}$$

The resultant air pollution index is scaled to relate to possible health effects, according to the standards of the Ontario Department of Energy and Resources. The Provincial Government advises industry and public institutions to curtail activities when the index reaches 32, but it cannot order major polluters to close down operations until the level reaches 50. A reading of 58 would greatly affect people suffering from severe respiratory diseases. If the level reaches 75, authorities can further curtail sources of pollution. An index reading of 100 would halt all non-essential sources because, at this point, the health of the general community would be affected. Compare this scale to examples of index readings calculated for the air pollution crises listed in Table 15.

TABLE 15

Location	Year	Calculated index reading
Osaka, Japan	1962	170
Detroit, U.S.A.	1952	175
New York, U.S.A.	1963	330
New York, U.S.A.	1962	440
London, England	1952	510

In each of these examples, the critical pollution levels persisted for at least three days. It would appear that Ontario cities have provided a wide measure of safety in dealing with potential air pollution episodes. However, recent studies suggest long-term health hazards from the average low quantities of pollutants present in urban atmospheres.

Occupational groups, such as asbestos handlers, are often studied in order to disclose possible health hazards and also to establish "safe" exposure levels for the many contaminants which find their way into the atmosphere. Such research is complicated by the fact that only those workers who develop a strong resistance to the particular irritants will remain in the given field of work. More sensitive individuals often must change occupations. Can industrial tolerance standards determined by testing hardier individuals be safely applied to an entire community? Not according to many medical authorities. Intensive research is necessary to evolve sensible air quality standards. And the battle has only begun. The stakes are high, for air pollution affects everyone. To quote a wise little character named Pogo—"We have met the enemy and He is Us!"

For Thought and Research

1 (a) Does your community have air quality standards or an air pollution index?

(b) If so, what air pollutants are monitored for this pollution index? Why?

(c) How are these particular pollutants monitored? How is the pollution index determined? You may have to consult local authorities for information.

(d) Are these air quality standards ever exceeded? If so, how often? Why?

(e) What measures are taken if these standards are exceeded? Is the present enforcement strict enough?

(f) Is a local authority available to whom you can report any infringements of local air pollution bylaws, for example, illegal incineration? Do you make an effort to report such incidents?

2 Are any workers in your community regularly exposed to high levels of airborne substances such as sulfur dioxide or asbestos? Are air quality standards enforced where these individuals work? If not, why not?

Recommended Readings

1 "Ontario's Air Pollution Index" by L. Shenfield and F. Frantisak, *Water and Pollution Control*, November, 1970. This article includes a more detailed explanation of the calculation of air pollution indices for several North American cities.

2 "The Control of Air Pollution" by A. J. Haagen-Smit, *Scientific American*, January, 1964.

3 *Air Pollution Index and Alert System*, Ontario Department of Energy and Resources Management. Similar papers for your area can likely be obtained from the nearest government branch in charge of pollution control.

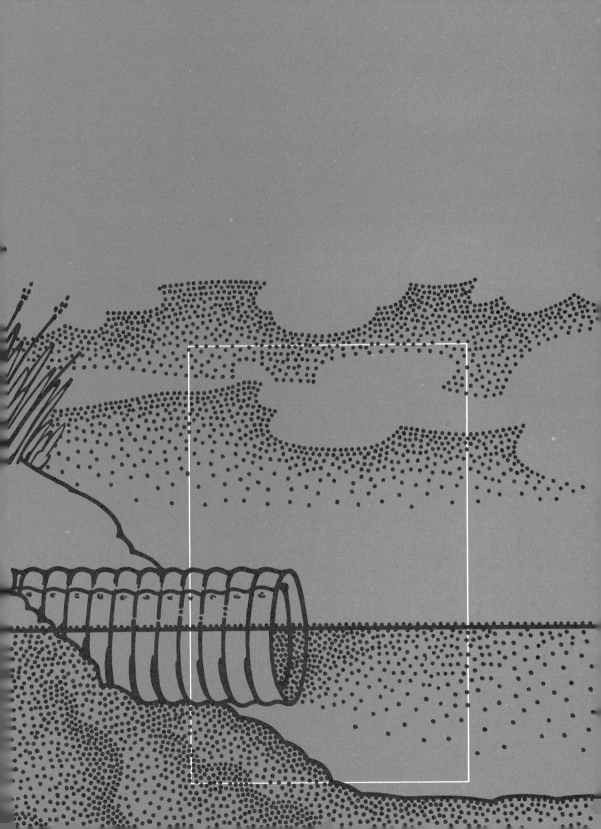

Research Topics in Pollution

5

This unit contains a number of topics in pollution that are of great importance to all North Americans today. Examine the topics and select one that interests you. Research it thoroughly, by yourself or with several other students.

A list of things that you might do has been included with each topic. This list will help you to get started on your project. It is *not* complete. Many other ideas will come to mind as you research the topic. Some references have been included. You should also consult encyclopedias and other books in your school and community libraries. Many government agencies, industries, and conservation groups supply free information on some of these topics. You need only ask for it. Letters, telephone calls, and visitations are effective ways of obtaining first-hand information.

Although you may not be affected directly by some of these forms of pollution, you are affected indirectly by all of them. For example, you may live in an area where noise pollution is not an obvious problem. But noise pollution lowers the efficiency of city workers which, in turn, affects the economy of the entire country, including your area. So think beyond yourself, beyond your community, beyond your province or state, and even beyond your country. Pollution is a global problem that requires global thinking.

At the end of your study you should prepare a written report so that others in your class can benefit from your work. Your teacher will give you instructions regarding the nature of the report.

5.1   THE POPULATION EXPLOSION

The human population in 6,000 B.C. was about five million people. By 1650 A.D. it was 500 million. By 1850 it had reached one billion. By 1930 it had risen to two billion. It is now rapidly approaching four billion. At one time, it took 200 years for the earth's population to double. Later it required only 80 years. Today the doubling time is about 35 years. An "explosion" is truly occurring.

As you might expect, the doubling times vary from country to country. In the so-called "underdeveloped countries" doubling times are around 20 to 30 years. For example, El Salvador has a doubling time of 19 years. Thus every 19 years El Salvador must double its production of food, its supply of housing and clothing, its roads, and so on if it wishes to maintain its present standard of living. Can El Salvador do this? Among "developed countries" doubling times range from 50 to 200 years. The United States, Russia, and Japan now have doubling times of about 65 years. That of the United Kingdom is 140 years and of Austria, 175 years.

If the current doubling time is maintained for the next 900 years, the earth will have 100 people on every square meter of its surface—land and water! But that's a long time from now, isn't it? Think for a moment, though, about the problems that Calcutta, India, will have to face in just 30 years when its estimated population will be 70,000,000. That's about nine times the present population of New York City. You are, no doubt, well aware of the problems that cities like New York already have. Will Calcutta, in the year 2000, be able to cope with problems many times greater than those of present-day New York?

Many people feel that science will find a solution to the population explosion without placing restrictions on population growth. They maintain, for example, that the earth can feed many times its present population. It probably can. But, at the moment, farmers in the United States and Canada are being paid *not to grow food* while close to four million people, mostly children, starve to death each year throughout the world. Suppose the earth *can* support 10 times its present population. Where are these people going to live? What problems will arise when people are packed closer and closer together? Can medical science handle the inevitable epidemics of diseases like cholera and typhoid? Where will these people deposit their wastes? Where will they get minerals, gasoline, and clean water? Where will they play?

Clearly the population explosion is the basic cause of environmental pollution. For example, water and air pollution were not problems until the world's increasing population put excessive strains on the ecosystem.

The biological solution to the population explosion is so simple that a seven year old can understand it—family size must be restricted to two children. But the implementation of this solution is another matter. At the moment, the social, economic, political, and religious obstacles seem almost insurmountable.

For Thought and Research

The following are some aspects to investigate:

1 Present population of the world; its distribution by continents and by major countries.

2 Rate of growth and projected rate of growth of the population of the world, the continents, and the major countries.

3 Rate of increase in the available food supply compared to the rate of increase in population; relationship of population growth to availability of good agricultural land.

4 The growth curve for the human population.

5 The growth curves of extinct animals and of animals that have stabilized populations.

6 Psychological effects of high population densities.

7 Other problems that accompany high population densities.

8 Factors that tend to keep animal populations in balance. Do the same factors affect human populations?

9 Relationship of the population explosion to the problem of environmental pollution.

10 Proposed solutions to the population explosion.

11 Factors that tend to hinder attempts to deal with the problem of population growth.

12 The dependence of our economy on population growth.

Recommended Readings

1 *The Population Bomb* by P. R. Ehrlich, Ballantine Books, 1968. This book is devoted to the scientific, political, social, and economic aspects of the population explosion. It is written in layman's language. It strongly emphasizes the need to immediately regulate population growth.

2 *Population, Resources, Environment* by P. R. Ehrlich and A. H. Ehrlich, W. H. Freeman, 1970. A thorough and more scientific treatment of the population explosion and associated problems by the same author as above.

3 *Concepts of Ecology* by E. J. Kormondy, Prentice-Hall, 1969. Consult this book for scientific information on the growth of populations.

The following books all contain articles on the population problem. In most cases you can tell by the title of the book the approach that the article will take.

4 *Pollution Probe* by D. A. Chant, New Press, 1970.

5 *Moment in the Sun* by R. Rienow and L. T. Rienow, Ballantine Books, 1967.

6 *Eco-Catastrophe* by the Editors of *Ramparts*, Harper & Row, 1970.

7 *The Environmental Handbook* by G. deBell, Ballantine Books, 1969.

8 *Man in the Web of Life* by J. H. Storer, The New American Library, 1968.

9 *Politics and Environment* by W. Anderson, Goodyear, 1970.

5.2   NOISE POLLUTION

Many experts estimate that the average noise level in North America is doubling every 10 years. Common sense tells us that, sooner or later, all of us will be affected by this noise. In many highly industrialized areas, it is already high enough to cause hearing impairment, nervousness, hypertension, ulcers, and suicidal tendencies. Clearly noise pollution is a problem that must be faced now.

For Thought and Research

The following are some aspects to investigate:

1 Measurement of noise using the decibel scale; threshold frequency; threshold of pain; intensity that produces hearing damage.

2 Effects of noise pollution on man and other animals. Can man adapt to increasingly higher noise levels?

3 The present magnitude of the problem; projected changes in the nature of the problem.

4 Chief sources of excessive noise. You might include here a study of one area of your community.

5 Laws regarding noise pollution. Who sets them? What are they? How strictly are they enforced?

6 Control of noise pollution. You might investigate some specific sources such as power lawnmowers, motorbikes, sports cars, transport trucks, snowmobiles, and motorboats. Are the high noise levels produced by these devices necessary? How could the noise be decreased? Is cost the main reason why noise has not been decreased? Are the noise levels below the limit set by law?

7 Is an ever-increasing noise level an inevitable consequence of our way of living? Would laws that prohibit excessive noise affect the economy in any way?

Recommended Readings

The following books contain articles that speak strongly against noise pollution. Be sure that you place emotionalism and scientific fact in proper perspective as you read these books.

1 *The Environment* by the Editors of *Fortune*, Harper & Row, 1970.
2 *Pollution Probe* by D. A. Chant, New Press, 1970.
3 *Moment in the Sun* by R. Rienow and L. T. Rienow, Ballantine Books, 1967.
4 *Man Against His Environment* by R. Rienow and L. T. Rienow, Ballantine Books, 1970.

5.3   SOLID WASTE DISPOSAL

Solid waste disposal is a problem closely tied to the population explosion. As cities grow, they either pollute the air by burning growing mounds of garbage or they truck the garbage farther and farther to disposal sites. Today the latter is becoming more common. Indeed, many cities are now making arrangements for the inevitable—transporting garbage by train to disposal sites hundreds of miles away. Is this the answer? Is moving garbage from your front yard to someone else's backyard the best that modern technology can do?

Each North American produces a little more than 5 pounds of solid wastes per day. This amounts to over 200 million tons per year for Canada and the United States. These figures do not include the millions of automobiles junked every year, nor do they include most industrial wastes. By 1980 these figures are expected to rise to 8 pounds per person per day, or a total of 350 million tons per year for the two countries. Where are we going to put it? Is it all really "garbage"?

For Thought and Research

The following are some aspects to investigate:
1 Methods used to dispose of municipal garbage; pros and cons of each method.
2 Problems facing municipalities now and in the future.
3 The method or methods used by your municipality; cost to the taxpayer. Visit a local incinerator or a landfill project, if possible.
4 Effects of solid wastes on the environment.
5 Specific problems, such as those created by non-returnable bottles and cans. Survey the density of these containers on a measured length of highway near your community. Has your state or province passed, or even considered, legislation on this problem? How have other areas handled the problem? Interview other students and your neighbors to learn the general consensus of opinion regarding non-returnable containers. Would a deposit of 25 cents per bottle insure the return of bottles to stores? What do you think the law should be about tossing bottles and cans onto the roadside? How would you have this law enforced?
6 Recycling of the components of garbage; its feasibility. Relationship of recycling to another problem—ever-decreasing natural resources.

7 Are economic problems associated with recycling? Would banning the sale of pop and beer in cans have any economic consequences? How strongly does our economy depend upon a "throw-away" style of living?

8 The population of the world is about 3 billion. Calculate the number of tons per year of solid wastes that would be produced if the entire world had our standard of living today. How many additional tons would be produced in 1980? What conclusions do you draw from these calculations?

Recommended Readings

1 *Cleaning Our Environment. The Chemical Basis for Action*, American Chemical Society, 1969. A source of accurate facts and figures. Read this source before reading the strongly written articles in the following two books.
2 *Politics and Environment* by W. Anderson, Goodyear, 1970.
3 *Man Against His Environment* by R. Rienow and L. T. Rienow, Ballantine Books, 1970.

5.4 **SEWAGE TREATMENT**

Sewage treatment, like solid waste disposal, becomes an increasing concern as our population grows. Improvements in sewage treatment must be made for two main reasons. First, inadequately treated sewage is rapidly destroying the waterways of North America. Second, our current method of sewage treatment has the effect of transferring nutrients from our productive farmlands to the oceans—you eat products from the farms; plant nutrients in your urine and feces go into the sewage system and, eventually, to the oceans. In countries like China that do not have extensive sewage systems, these nutrients are, for the most part, returned to the land. Can the plant nutrients in our sewage be reclaimed? Can the quality of our sewage treatment be improved?

For Thought and Research

The following are some aspects to investigate:
1 Primary treatment and secondary treatment. How effective is each method in removing suspended solids and dissolved solids? How do they work? (Use diagrams.)
2 Advanced or tertiary treatment; the different types and what they do. (Use diagrams.) Recycling of waste water; can it be done? Cost?
3 Problems associated with municipal sewage treatment; cost per gallon for each type of treatment.
4 What kind of sewage treatment facilities does your community have? Is your community making plans for better treatment of its sewage? Does your community

have advanced treatment methods to remove nutrients like phosphorus? What are the average T.D.S. and T.S.S. values of the effluent? What is the average B.O.D. of the effluent? A visit to your local sewage treatment plant would help here. The plant manager should have these figures. Do your own T.D.S. and T.S.S. determinations as outlined in Unit 6. How do your values compare with the values you obtained from the plant manager? If there is a significant difference between your values and the average values, what might be the reason for the difference?

5 What level of government sets the standards for water quality in your area? Who do you think should set the standards? Why? Who is responsible for enforcing the law? Are municipalities forced to obey the same law as industries?

6 Are there any rivers or lakes near your community that are clearly polluted because of municipal sewage? If the answer is "yes," find answers to these questions:

(a) Are warning signs posted?

(b) If there are no warning signs, who is responsible for putting them up? Why were they not put up?

(c) Is your community breaking the law by polluting the body of water?

7 You may want to research the methods used to purify water for drinking purposes. A visit to a water purification plant would be helpful. What methods does your community use? Where does the water come from? Are there any particular problems that add to the difficulty of purifying the water?

8 If you live in the Great Lakes region, you should investigate the work of the International Joint Commission of Canada and the United States which is attempting to establish standards of water quality for the lakes. A Canadian government official recently accused the United States government of not moving rapidly enough in cleaning up sewage entering the Great Lakes from cities in the United States. A United States government official replied that difficulties were encountered because of a different political situation in the U.S. What does he mean? Can this situation be changed?

Recommended Readings

The first two references contain factual information that will be helpful to you in researching this topic. The third reference gives the opinions of persons who have studied the matter carefully.

1 *Cleaning Our Environment. The Chemical Basis for Action*, American Chemical Society, 1969.

2 *The Biological Aspects of Water Pollution* by C. G. Wilber, C. C. Thomas, 1969.

3 *Pollution Probe* by D. A. Chant, New Press, 1970.

5.5   PESTICIDES IN THE ENVIRONMENT

All of us carry pesticides in our body tissues. Many species of animals are in danger of becoming extinct because of the accumulation of pesticides in food chains; and man is part of many food chains.

Recently great concern has been expressed in many quarters over the indiscriminate use of pesticides. What is indiscriminate use? Does a city dweller need to spray his lawn several times a year with herbicides and insecticides? Should a fruit farmer spray his orchard with deadly chemicals before any sign of infection can be seen? Clearly all of us need a greater awareness of the nature, uses, and effects of pesticides if the concern over them is going to amount to anything more than talk.

For Thought and Research

The following are some aspects to investigate:

1 What is a pesticide?
2 Some commonly used fungicides, herbicides, and insecticides. Specifically, why are they used?
3 The magnitude of the problem.
4 Are pesticides necessary?
5 The most dangerous pesticides; the persistent pesticides; accumulation in food chains.
6 Effects on birds, fish, and other animals.
7 Effects on human health.
8 Laws regarding the use of DDT and other persistent chemicals.
9 Can contamination of the environment with pesticides be reduced without harming the economy? How closely does farm productivity depend on the use of pesticides? Could the world feed itself without using pesticides?
10 Alternative methods of pest control.
11 The use of pesticides by the home gardener. A survey of the types used in your neighborhood would be of interest. How many of the people interviewed use pesticides on a regular basis, as a preventive rather than a corrective measure? Under what circumstances do you think home gardeners should use pesticides?
12 Many countries have virtually eliminated malaria by using DDT to kill the mosquitoes that carry the disease. If DDT were banned in those countries, millions would die each year from malaria. Yet, every living organism on the earth is gradually accumulating this DDT. What are its effects? Less persistent pesticides exist that kill the mosquitoes but they are too expensive for the "under-developed countries" to use. Do the United States and Canada have a moral obligation to supply such pesticides?
13 The use of DDT has been virtually banned over most of North America. In some localities, however, limited use is allowed. For example, in Ontario tobacco farmers can and do obtain permits to use DDT. What do you think of this?

Recommended Readings

For factual, scientific information consult:

1 *Cleaning Our Environment. The Chemical Basis for Action*, American Chemical Society, 1969.
2 *The Biological Aspects of Water Pollution* by C. G. Wilber, C. C. Thomas, 1969.

3 *Readings in Conservation Ecology* by G. W. Cox, Meredith, 1969.

For articles that combine a minimum of scientific detail with debates regarding the use of pesticides, see:

4 *Pollution Probe* by D. A. Chant, New Press, 1970.
5 *Moment in the Sun* by R. Rienow and L. T. Rienow, Ballantine Books, 1967.
6 *The Environmental Handbook* by G. deBell, Ballantine Books, 1969.
7 *Politics and Environment* by W. Anderson, Goodyear, 1970.
8 *Man in the Web of Life* by J. H. Storer, The New American Library, 1968.
9 *Man Against His Environment* by R. Rienow and L. T. Rienow, Ballantine Books, 1970.
10 *Population, Resources, Environment* by P. R. Ehrlich and A. H. Ehrlich, W. H. Freeman, 1970.

The following two books are devoted entirely to the pesticide issue:

11 *Silent Spring* by R. Carson, Houghton Mifflin, 1962.
12 *Since Silent Spring* by F. Graham, Houghton Mifflin, 1970.

5.6 THERMAL POLLUTION

As the demand for electric power increases, more generating plants must be built. These generating plants use natural bodies of water for cooling purposes; hence they are among the main sources of thermal pollution. Many industries also use natural waters as a coolant. Thus one result of more people and a higher standard of living appears to be warmer water. Very small changes in temperature have marked effects on most living organisms in aquatic ecosystems.

For Thought and Research

The following are some aspects to investigate:

1 The main sources of thermal pollution.
2 Present seriousness of the problem; projections for the future.
3 Effects of thermal pollution on aquatic life. After you have researched this aspect, you could conduct an investigation of the effects on aquatic life of hot effluent from a nearby hydro generating plant or industry.
4 Ultimate effects on the world ecosystem.
5 Methods of minimizing thermal pollution.
6 Power companies are undoubtedly aware of the consequences of thermal pollution, yet many of them have active advertising campaigns that encourage the consumer to "live better electrically" by using more electricity. In some cases this is true even though the company is government-owned and operated. Debate this policy.
7 Would a cut-back in the use of electricity be a reasonable alternative to the building of more power plants? For example, could neon street signs, electric brooms, electric can-openers, and air conditioners be banned? Do the affluent have

the right to use such devices when, ultimately, all people will be affected? Could the North American economy sustain itself if non-essential electrical devices were banned?

Recommended Readings

All of these books contain information on thermal pollution. The first contains reliable information on biological aspects. The other three, combined, will fill in your knowledge in the political, social, and economic areas.

1 *The Biological Aspects of Water Pollution* by C. G. Wilber, C. C. Thomas, 1969.
2 *Pollution Probe* by D. A. Chant, New Press, 1970.
3 *Man Against His Environment* by R. Rienow and L. T. Rienow, Ballantine Books, 1970.
4 *Population, Resources, Environment* by P. R. Ehrlich and A. H. Ehrlich, W. H. Freeman, 1970.

5.7   RADIATION POLLUTION

As the demand for electrical power increases, more nuclear power plants will be built. Radioactivity escapes from these plants into the surroundings. Also spent fuel from the plants is still radioactive when it is disposed of. Hospitals and some research laboratories use radioactive materials and must dispose of radioactive residues.

Radiation pollution is not considered by most people to be a problem at the moment. The indications are, however, that it could become a serious problem in the near future as the pressures of increasing population make necessary the use of more and more radioactive materials.

For Thought and Research

The following are some aspects to investigate:
1 Sources of radioactive wastes.
2 Present seriousness of the problem; projections for the future.
3 Possible effects on humans of drinking water that is contaminated with radioactive material.
4 Other ways that radioactive materials can come in contact with humans.
5 Accumulation of radioactive materials in food chains. Suggest possible food chains that radioactive substances might follow in traveling from an industrial effluent to humans.
6 Permissible levels of radioactivity.
7 Effects on organisms other than man.
8 Methods used to minimize radiation pollution.

Recommended Readings

For scientific information on sources and effects of radiation pollution consult:

1 *The Biological Aspects of Water Pollution* by C. G. Wilber, C. C. Thomas, 1969.

2 *Readings in Conservation Ecology* by G. W. Cox, Meredith, 1969.

3 *Ionizing Radiation and Life* by T. G. Overmire, B.S.C.S. Patterns of Life Series, Rand McNally, 1970.

The following books also contain information on radiation pollution and, in most cases, discuss political, social, and economic aspects of the problem:

4 *Population, Resources, Environment* by P. R. Ehrlich and A. H. Ehrlich, W. H. Freeman, 1970.

5 *Pollution Probe* by D. A. Chant, New Press, 1970.

6 *Eco-Catastrophe* by the Editors of *Ramparts*, Harper & Row, 1970.

7 *Man Against His Environment* by R. Rienow and L. T. Rienow, Ballantine Books, 1970.

8 *Politics and Environment* by W. Anderson, Goodyear, 1970.

5.8   OIL POLLUTION

Another consequence of the growing population of the earth is the demand for more crude oil. Fuels, chemicals, synthetic fabrics, plastic containers, and many other things are made from it. As accessible sources of crude oil become exhausted, the oil companies move out to sea and up to the Arctic to tap new reserves. Accidents during the transporting of oil over long distances are inevitable. The consequences of such accidents are disastrous.

The effluents from oil refineries pollute bodies of water. The water used as a coolant in the refineries creates a thermal pollution problem. Also, most refineries are notorious air polluters.

What do we do? Do we lower our standard of living so that we do not require as much oil? Do we pay more for the products and insist that pollution stop? Do we rely on the refinery owners to end pollution because they, too, are ultimately going to be affected?

For Thought and Research

The following are some aspects to investigate:

1 Causes of oil pollution: oil spills, off-shore drilling, refinery effluents.

2 Effects of refined oil, crude oil, and refinery effluents on the environment; effects on aquatic life and waterfowl.

3 Look into the causes, effects, and cleanup procedures involved in a major oil spill such as the Torrey Canyon disaster or the eruption in the Santa Barbara Channel.

4 Special problems associated with oil spills in the Arctic regions of Canada and Alaska; super-tankers; ecological effects of oil pipelines on the tundra.

5 How dependent is our standard of living on crude oil?

6 Should off-shore drilling be banned? Why?

7 Plans made by the government to deal with future oil spills.

8 International cooperation to deal with possible oil spills in the Great Lakes or in coastal waters.

9 Is the competition among oil companies apt to create more or less pollution?

Recommended Readings

1 *The Biological Aspects of Water Pollution* by C. G. Wilber, C. C. Thomas, 1969. Contains an excellent chapter on the causes and effects of oil pollution.

2 *Pollution Probe* by D. A. Chant, New Press, 1970. Discusses briefly the effects of some of the major oil spills that have occurred.

3 *Eco-Catastrophe* by the Editors of *Ramparts*, Harper & Row, 1970. Contains an article on the Santa Barbara oil well eruption and one on the special problems of oil pollution in Alaska. Political aspects are discussed.

5.9   PULP MILL POLLUTION

The pulp and paper industry is one of the largest in North America. Large amounts of water are used in a pulp and paper operation. The water is commonly returned to a lake or river with added pollutants. Since most pulp and paper mills were initially built far from large centers of population, the seriousness of their contribution to water pollution was not recognized until recently. Modern technology has made this pollution unnecessary. Indeed, some mills have already cleaned up their operations. They either recycle the water or purify it considerably before releasing it. Other plants continue to pollute.

For Thought and Research

The following are some aspects to investigate:

1 What goes on in a pulp and paper mill?

2 In general, what materials are found in the effluent from such a plant? What harm does each one of these do to an aquatic ecosystem?

3 Is complete purification of the effluent possible?

4 The magnitude of the problem.

5 Visit a pulp and paper mill, if possible. Examine the river or lake that receives the effluent. Ask your tour guide to explain the company's plans for pollution abatement. Test the water quality at suitable sites using the appropriate tests from Unit 6.

6 If you cannot visit a pulp and paper mill, write for literature that explains the plant operation. Ask particularly for information regarding pollution control measures.

7 Would recycling of paper and paper products lessen pollution? Would people be put out of work if recycling became widespread?

8 The economy of many small towns depends almost entirely upon local pulp and paper operations that employ many of the town residents. How do you think this might affect the chances of a pollution abatement program being initiated by the paper mills?

Recommended Readings

1 *The Biological Aspects of Water Pollution* by C. G. Wilber, C. C. Thomas, 1969. Contains a thorough discussion of types of pollutants from pulp and paper mills and the effects of these pollutants on living organisms.

2 *Pollution Probe* by D. A. Chant, New Press, 1970. Describes some of the effects of water pollution by pulp and paper operations and deals with some of the political issues involved.

5.10 LAND USE AND THE MANAGEMENT OF NATURAL RESOURCES

The earth is a finite system. Thus its supply of resources such as crude oil and minerals is fixed. However, the demand for these resources is not fixed. It is increasing at an incredible rate, particularly in the "developed countries." For example, in 1970 the United States consumed close to 50% of the non-renewable resources that were used in the world during that year. Yet the United States contains only 6% of the world's population. If the demand for crude oil and minerals continues to grow as fast as experts have estimated, the United States will require 100% of the world's non-renewable resources by 1980. That leaves nothing for the rest of the world!

Our forests, fresh water, and land must also be considered as natural resources. Would you call these renewable or non-renewable resources?

For Thought and Research

The following are some aspects to investigate:
1 Relationship to pollution problems.
2 Relationship to the population explosion.

3 On what kind of land is your town or city built—prime agricultural land, marginal agricultural land, or poor agricultural land? Why? What kind of land is being consumed as your municipality grows? Why? Does the location of your town or city contribute to air or water pollution?

4 Are your city parks designed in such a way that automobiles can cruise back and forth in the parks? Do your parks assist in the abatement of air pollution?

5 Are forests being replaced as fast as they are being cut down? Can paper be recycled? Are forests necessary?

6 Can "tin" cans and car bodies be recycled? Are they? Would recycling help to lower air and water pollution?

7 What is the policy of your state or provincial government regarding the preservation of some forest areas in their natural condition?

8 What are the projected "life-times" of the world's supply of important natural resources such as crude oil, iron, copper, nickel, and uranium? What proportion of these resources does North America consume? How does this relate to the ratio of the population of North America to the population of the rest of the world? What relationship exists between rate of consumption of natural resources and pollution?

9 Interview other students and your neighbors regarding suggestions like this one: "Snowmobiles, all-terrain vehicles, trail bikes, and excessively large cars should be banned to conserve natural resources and to lower pollution in cities and in the countryside."

10 Do "developed countries" have the right to consume non-renewable resources at a rate that far exceeds that of "under-developed countries"?

11 What do you think would happen if the United States decided that the percentage of the world's non-renewable resources consumed by a country should equal the percentage of the world's population that lives in that country? Do you think that this is a reasonable proposal? Why?

12 Do you think that it is possible for all countries to attain a standard of living comparable to ours? Why?

Recommended Readings

All of these books contain articles on land use and natural resources:

1 *The Environment* by the Editors of *Fortune*, Harper & Row, 1970.
2 *Pollution Probe* by D. A. Chant, New Press, 1970.
3 *Moment in the Sun* by R. Rienow and L. T. Rienow, Ballantine Books, 1967.
4 *Man in the Web of Life* by J. H. Storer, The New American Library, 1968.
5 *The Ecological Conscience* by R. Disch, Prentice-Hall, 1970.
6 *The Environmental Handbook* by G. deBell, Ballantine Books, 1969.
7 *Eco-Catastrophe* by the Editors of *Ramparts*, Harper & Row, 1970.
8 *Politics and Environment* by W. Anderson, Goodyear, 1970.
9 *Population, Resources, Environment* by P. R. Ehrlich and A. H. Ehrlich, W. H. Freeman, 1970.

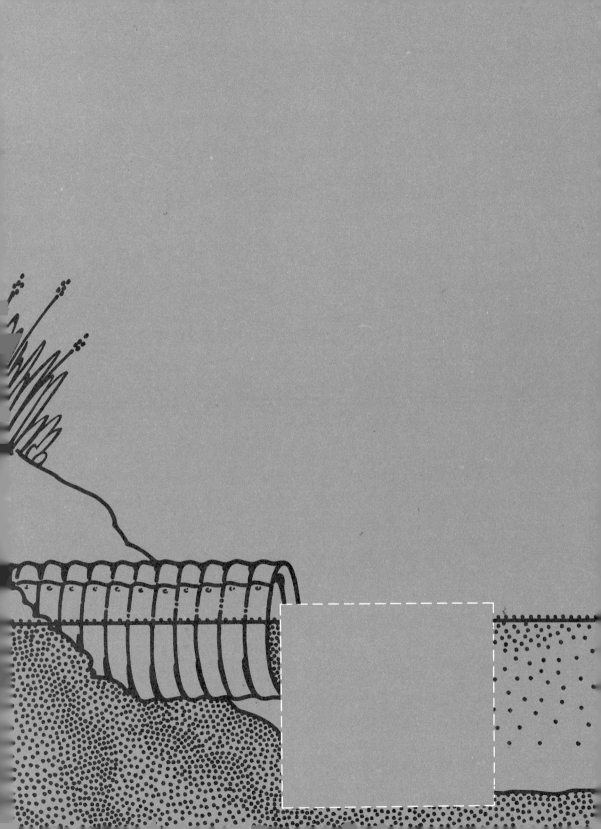

Field and Laboratory Studies in Water Pollution

6

This unit and the next consist of experiments, laboratory exercises, field techniques, and field studies. It is not intended that you do everything in these units. Whether or not you do some of these studies depends on the equipment available and on the type of pollution study that you wish to perform. You should, however, do all of the experiments and laboratory exercises that are mentioned in the first four units. You should also try in the classroom any techniques that you intend to use in the field. Get to know the equipment and procedures so that you can perform the tests quickly and accurately once you are in the field. For example, if your study of a polluted stream calls for a D.O. test, try the test on aquarium water before you go on the field trip.

6.1 THE ROLE OF PRODUCERS IN ECOSYSTEMS

In Section 1.4 you were introduced to the process of photosynthesis:

Carbon Dioxide + Water + Light Energy

$$\xrightarrow{\text{chlorophyll}} \text{Glucose} + \text{Oxygen}$$

$$6\ CO_2 + 6\ H_2O + \text{Light Energy}$$

$$\xrightarrow{\text{chlorophyll}} C_6H_{12}O_6 + 6\ O_2$$

Glucose belongs to a family of organic compounds called carbohydrates. If you write its formula as $C_6(H_2O)_6$ you can probably see why it is called a carbohydrate. What products do you think will be formed if some glucose is heated in a test tube? Try it. Do you see why it is called a carbohydrate?

Plants convert much of the glucose that they produce into starch, another carbohydrate. Heat some starch in a test tube and compare the products formed with those obtained when you heated glucose.

Glucose is difficult to detect in plants. However, starch is easily detected. Make a paste of starch and water on a glass plate. Add a few drops of iodine solution. The color that you observe will appear even if only a few ppm of starch are present. This *starch test* is used to test for the presence of carbohydrates in some of the experiments in this section.

A. THE PRODUCTION OF CARBOHYDRATES
   BY PRODUCERS

Materials

a) 150 ml beaker
b) 600 ml beaker
c) hot plate
d) forceps or crucible tongs
e) glass plate
f) ethyl or isopropyl alcohol
g) geranium plant that has been under a light for 24 hours
h) geranium plant that has been in the dark for 24 hours

Procedure

a) Remove a leaf from each geranium plant.

b) Immerse the leaves in a beaker of boiling water for a few seconds. Remove them as soon as they become limp.

c) Transfer the leaves to a beaker containing boiling alcohol. Use a hot plate or place a small beaker of alcohol in a larger beaker partly full of boiling water. DO NOT HEAT THE ALCOHOL WITH AN OPEN FLAME. IT WILL IGNITE.

d) When the leaves are white, remove them from the alcohol. Soften them by dipping them in boiling water for one or two seconds.

e) Spread the leaves on a glass plate and cover them with iodine solution.

f) Record your observations.

Discussion

What do these observations tell you about the role of geranium plants in the ecosystem in which they are found?

Repeat the experiment using the leaves of corn seedlings, bean seedlings, and other plants. What do these observations tell you about the role of producers in ecosystems?

B. CHLOROPHYLL AND PHOTOSYNTHESIS

Is chlorophyll necessary for photosynthesis? Many plants have leaves that contain chlorophyll in some regions and not in others. Examples are variegated geraniums and some coleus plants. Obtain one of these plants and design an experiment to answer the question. You will require many of the materials and procedures of part A.

C. LIGHT AND PHOTOSYNTHESIS

Is light required for photosynthesis? You may think that this question was answered in part A. However, you used two different plants in that experiment. A scientist would say that you did not have very good *controls* in the experiment. The different results with each plant could be entirely due to the fact that they were different plants. To eliminate this possibility, you should use only one plant or, preferably, one leaf of that plant. See if you can design such an experiment. Again, you will require many of the materials and procedures of part A.

What do these results tell you about the role of producers in ecosystems?

D. CARBON DIOXIDE AND PHOTOSYNTHESIS

Does a green plant use carbon dioxide during photosynthesis?

Materials

a) bromthymol blue
b) carbon dioxide
c) glass tube or soda straw
d) *Elodea*

e)   test tubes

f)   light source

Procedure

a)   Bromthymol blue is an acid-base indicator. Find out what color it turns in an acidic solution; put 3 or 4 drops of bromthymol blue in a test tube of water that contains a few drops of acid. Perform a similar experiment to find out what color bromthymol blue turns in a basic solution.

b)   What is formed when carbon dioxide comes in contact with water? (See Section 2.2.) Confirm this by bubbling carbon dioxide from a gas cylinder or from your breath into a neutral solution of bromthymol blue.

c)   Design an experiment to show whether or not green plants use carbon dioxide during photosynthesis. You will need all of the materials listed above. Don't forget that a control is necessary.

d)   What do these results tell you about the role of producers in ecosystems?

E. OXYGEN AND PHOTOSYNTHESIS

Do green plants produce more oxygen during photosynthesis than they require for respiration?

Materials

a)   2 test tubes

b)   2 1000 ml beakers

c)   2 funnels

d)   *Elodea*

e)   sodium bicarbonate

Procedure

a)   Set up the apparatus shown in Figure 6-1. The water must contain carbon dioxide during the entire experiment. To insure this, add 2 or 3 pinches of sodium bicarbonate to the water. Insert the cut ends of the *Elodea* sprigs into the stem of the funnel. The funnel should be deep in the beaker of water. The test tube and funnel

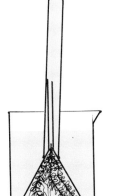

Fig. 6-1
Oxygen and photosyn-
thesis.

must be full of water at the beginning of the experiment. You figure out how to fill them.

b) Set up a control experiment.

c) Shine a bright light on the entire setup for several days. Watch for a product. If one appears, confirm its identity with a suitable test.

d) What does this experiment tell you about the role of producers in ecosystems?

F. CARBON DIOXIDE AND RESPIRATION

Do green plants produce carbon dioxide? Green plants, like all living organisms, respire. Was there any evidence in the experiment of part D that green plants give off carbon dioxide? Account for your answer.

Design an experiment to show that a green plant produces carbon dioxide. A simple modification of part D should be sufficient. Again, don't forget to set up a control.

What does this experiment tell you about the role of producers in ecosystems?

6.2 THE ROLE OF DECOMPOSERS IN ECOSYSTEMS

This exercise illustrates the action of a common decomposer, yeast. Under the conditions described, yeast cells decompose sugar anaerobically. The process is called fermentation.

Materials

a) 4 fermentation tubes

b) 0.25M, 0.50M, 0.75M, and 1.00M solutions of sucrose (common sugar)

c) 1 package of dried baker's yeast

d) distilled water

e) Mohr pipet

f) 5 100 ml beakers

g) 50 ml graduated cylinder

Procedure

a) Prepare a suspension of yeast by thoroughly mixing 1.00 gm of yeast with 50 ml of distilled water.

b)   Number the beakers from 1 to 5. Add liquids to them as follows:

Beaker 1   40 ml of distilled water
Beaker 2   40 ml of 0.25M sucrose solution
Beaker 3   40 ml of 0.50M sucrose solution
Beaker 4   40 ml of 0.75M sucrose solution
Beaker 5   40 ml of 1.00M sucrose solution

c)   To each beaker add 2.0 ml of yeast suspension. Mix thoroughly.

d)   Number the fermentation tubes from 1 to 5. Transfer each mixture to the appropriate fermentation tube. Incline the tubes to remove all air from the closed arms.

e)   Place the tubes in a warm place (25–30°C). Examine the tubes every 2–3 hours. Note the volume of gas produced (Fig. 6-2). If the fermentation tubes are not graduated, mark the position of the gas at each time interval with a marking pen. At the end of the experiment, you can use these marks to determine the corresponding volumes. How?

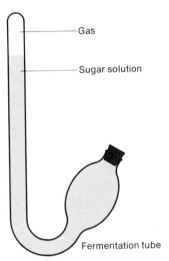

Fig. 6-2
The decomposing action of yeast on common sugar.

Gas

Sugar solution

Fermentation tube

f)   Continue the experiment for 3 or 4 days.

g)   Plot graphs of volume versus time for each mixture.

Discussion

What purpose is served by the mixture in tube 1? What gas do you think is formed? Test your hypothesis. What does this exer-

cise tell you about the role of decomposer organisms in ecosystems?

6.3 BACTERIA IN WATER: COLIFORM COUNTS

Two methods for determining the coliform count of water are
outlined here.

A. THE HACH COLIVER METHOD (Fig. 6-3)

The method is an adaptation of the American Public Health
Association (A.P.H.A.) test for coliform bacteria.

Materials

a) 5 presumptive tube assemblies (contain Lactose Broth)

b) 5 confirmation tube assemblies (contain Brilliant Green
Lactose Bile Broth)

Procedure

Presumptive Test

a) Wash your hands thoroughly with soap and water.

b) Remove a presumptive tube assembly from its sealed
package.

Fig. 6-3
The Hach Coliver method
for coliform testing.

c)   Remove the cap. Do not touch the inside of the cap or tube assembly. If you do, bacteria from your hands may enter the assembly.

d)   Fill the tube assembly with the water to be tested and replace the cap.

e)   Allow the nutrient medium to dissolve and then invert the assembly to fill the inner tube with water. Make sure that no air bubbles are in the inner tube.

f)   Repeat these steps with 4 other presumptive tube assemblies.

g)   Place the 5 tube assemblies in a warm place, preferably in an incubator at $35° \pm 0.5°C$. After 1 hour invert the tubes again. This removes from the inner tubes any air that the heat may have driven out of the water. Return the tube assemblies to the incubator.

h)   If a gas collects in the inner tube after 12 to 24 hours, coliform bacteria are presumed to be present. If no gas collects, return the samples to the incubator. Leave them there for a total of 48 hours. If no gas forms in this time, coliforms are presumed to be absent.

Confirmation Test

Since a few bacteria other than coliforms can produce gas with the Lactose Broth in the presumptive tubes, it is best to perform also the confirmation test. Production of a gas in a confirmation tube assembly verifies the presence of coliform bacteria.

a)   After gas has been observed in a presumptive tube assembly, invert the tube. This transfers some bacteria to the inside of the cap. Place the cap on a confirmation tube assembly.

b)   Invert the confirmation tube assembly to remove air from the inner tube.

c)   Repeat a) and b) for all presumptive tube assemblies that contain gas.

d)   Incubate the confirmation tube assemblies at $35° \pm 0.5°C$ for 48 ± 3 hours. The formation of a gas confirms the presence of coliform organisms.

Discussion

If you use 5 tubes as outlined in the procedure, you can estimate the probable number of coliform bacteria present with Table 16.

TABLE 16

| Number of tubes containing | | Coliforms per 100 ml |
No gas	Gas	
0	5	over 16
1	4	16
2	3	9.2
3	2	5.1
4	1	2.2
5	0	none

B. MEMBRANE FILTRATION: THE MILLIPORE METHOD (Fig. 6-4)

Membrane filters are thin, porous plaster screens. They are produced with a pore size so small that they can trap on their surface all of the bacteria in a water sample when the water is forced through. After a given volume of water is forced through, the filter is placed on a suitable nutrient medium. Individual bacteria then multiply in numbers to form visually identifiable colonies. Each colony that forms represents a single bacterium in the original water sample.

The Millipore test kits have instructions for total bacteria and total coliform counts. In addition, media can be purchased that will culture only *E. coli*. Remember that when bacteria are

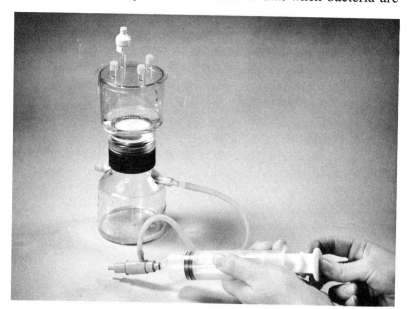

Fig. 6-4
A water sample being filtered using the Millipore method. (Courtesy of Millipore Corp.)

numerous in water, the sample should be greatly diluted with sterilized water before filtering. It is much easier to count 10 colonies from 1 ml of sample water diluted in 99 ml of sterile water than 1,000 colonies from 100 ml of sample water.

If you do not have access to Millipore equipment, you can grow bacterial colonies directly on an agar nutrient medium in a petri dish. Quantitative work, however, is nearly impossible without the membrane filter. *Do not, under any circumstances, begin to grow such cultures until you have carefully studied the safety techniques that are outlined in a microbiology laboratory book.* The directions for the selection of the appropriate medium and the techniques for starting and maintaining the culture can be found in the same book.

6.4 PRIMARY PRODUCTION

The primary production of an aquatic ecosystem is the rate at which energy from the environment is utilized to form organic compounds through photosynthesis. To arrive at an index of this productivity, the amount of oxygen produced by part of the system is measured. Why is the oxygen measured?

This is an *index* only. The actual process of photosynthesis varies with the rate of respiration of organisms, light intensity, photoperiod, cloud cover, seasons, turbidity, temperature, and other factors.

Materials

a)   500 ml bottles

b)   aluminum foil

c)   D.O. test kit

Procedure

a)   Determine the D.O. of a lake at various places and depths. Record the values.

b)   Collect two bottles of water at each of these locations. If it is rich in phytoplankton and zooplankton, the water is enough. Otherwise add some sprigs of larger aquatic vegetation to each bottle. Try to keep these sprigs fairly equal in size. Seal both bottles.

c)   Cover one bottle completely with aluminum foil.

d)   Suspend both bottles at the place and depth where they were taken. In the uncovered bottle, respiration and photosynthesis will continue. In the covered bottle, only respiration will occur.

e) After a period of time (1 to 24 hours), analyze each bottle for D.O. Record the values. The total oxygen produced for the time interval is the sum of the difference between the D.O. content at the start and the finish in each bottle. For example:

D.O. at end, uncovered	12.3 ppm
D.O. at start, uncovered	−6.4 ppm
Net oxygen produced	5.9 ppm
D.O. at start, covered	6.4 ppm
D.O. at end, covered	−1.9 ppm
Oxygen consumed	4.5 ppm
Net oxygen produced	5.9 ppm
Oxygen consumed	+4.5 ppm
Total oxygen produced	10.4 ppm

Thus an index of the productivity is the production of 10.4 ppm of oxygen in x units of time (or 10.4 mg of oxygen per liter of water in x hours).

6.5 ALGAE IN WATER

This study should be performed after you have read Section 3.5. Collect algal samples from as many of the following sources as possible: classroom aquaria, roadside pools, ponds, lakes (large and small; clean and polluted), streams (clean and polluted), near a sewage inlet, downstream from a sewage inlet.

Use a plankton net to concentrate the algae from ponds, lakes, and streams. In some cases you will have to scrape the algae from rocks. Be sure that the samples have air as they are being transported to the classroom.

Examine portions of each sample with a microscope. Try to identify the 2 or 3 most common species in each sample. Do your discoveries agree with what you expected to find? If not, why not? Of the zooplankton that you saw, which are primary consumers that feed on the algae? Are algae, alone, good pollution indicators? Explain your answer.

6.6 POLLUTION STUDIES USING A MINI-ECOSYSTEM

Collect several gallons of representative pond water. Get some surface water (including duckweed and the organisms that reside on and amongst it), some water from different depths, and, most important of all, some water from the space just above the bottom detritus. Include also 3 or 4 cups of the bottom detritus. Do not include macroscopic animals like fish or frogs. In such a confined volume, no balance could possibly be attained with such animals present. Do not reject, however, such animals as nematodes, tubificid worms, bloodworms, and planaria. A few sprigs of aquatic plants like *Chara* and *Elodea* can be included.

In the laboratory, divide this water and its contents as equally as you can among several containers. Each should have a capacity of at least 2 quarts. The number of containers depends upon the number of experiments that you intend to perform.

Select one of these as the control. Duplicate in it as many of the features of the natural pond as you can. For example, it should not be aerated. The light quality and intensity should correspond to the average light conditions in the natural setting. The life in this control should be examined every few days. Keep track of the species present and determine the relative sizes of the populations of the various organisms.

For each of the remaining containers of pond water, change one environmental factor. Monitor the effects that this change has on the life in the water. Try

a) a change in light intensity;

b) a change in D.O. concentration (achieved by aerating the water);

c) an increase in the concentration of a plant nutrient like phosphorus or nitrogen;

d) the addition of lawn fertilizer;

e) the addition of various types of detergents;

f) an increase in average temperature;

g) the addition of organic matter.

Don't attempt to do all of these. Also, don't expect instant results. In some cases no obvious changes will occur for several weeks or even for several months. This is a long-term experiment. Try to predict in advance the effects of the changes that you impose on the environment.

6.7 DISSOLVED OXYGEN (D.O.)

The test for dissolved oxygen (D.O.) is probably the most important test for determining water quality. The best and easiest methods for D.O. analysis use D.O. testing kits (Hach or La-Motte) or a D.O. meter. The kits use pre-weighed and pre-measured amounts of chemicals which are added to a water sample in a certain order. Titration is done on a direct-count basis or with a micro-buret. Oxygen meters are generally more accurate than kits, but are considerably more expensive and are temperamental.

Most kits are based on some modification of the Winkler method. If you do not have kits, you can use the modified Winkler test below. It is accurate to $\pm$ 0.5 ppm.

Materials

a)   250–300 ml bottles with ground glass stoppers (B.O.D. bottles)

b)   500 ml and 1 liter Erlenmeyer flasks

c)   100 ml volumetric pipet

d)   2 10 ml Mohr pipets

e)   buret

f)   manganese(II) sulfate solution: 480 gm of $MnSO_4 \cdot 4H_2O$, 400 gm of $MnSO_4 \cdot 2H_2O$, or 364 gm of $MnSO_4 \cdot H_2O$ dissolved in enough distilled water to make 1 liter of solution.

g)   alkaline-iodide solution: 500 gm of sodium hydroxide (or 700 gm of potassium hydroxide) and 150 gm of potassium iodide (or 135 gm of sodium iodide) dissolved in enough distilled water to make 1 liter of solution.

h)   concentrated sulfuric acid

i)   starch solution: 5 gm of soluble starch dissolved in 1 liter of distilled water. Sterilize the solution and store it in small bottles, to be opened as they are needed. This prevents rapid deterioration of the solution.

j)   sodium thiosulfate solution: dissolve 3.953 gm of $Na_2S_2O_3$ in enough distilled water to make 1 liter of 0.025M solution. Add 5 ml of chloroform to preserve it. A new solution should be made up every 3 to 4 weeks. This solution must be standardized occasionally with 0.025M potassium dichromate as described below.

Standardizing the Sodium Thiosulfate Solution

a)   Dissolve 2.5 gm of KI in 50 ml of distilled water.

b)   Add 0.5 ml of concentrated sulfuric acid.

c)   Add 20 ml of 0.025M potassium dichromate solution (6.96 gm per liter of solution).

d)   Place the solution in a dark place for 5 minutes.

e)   Titrate this solution with the sodium thiosulfate solution to be standardized, using a few drops of starch solution as the indicator. If the sodium thiosulfate solution is 0.025M, only 20 ml are required to reach the end point. To find the correction values, use the following rules:

If more than 20 ml were required, then the solution is weaker than 0.025M and the correction value should be less than 1. For example, if 21 ml were used, divide 20 by 21 to give 0.95, the correction factor. Thus, if 13 ml of sodium thiosulfate solution are used in a D.O. titration, the corrected D.O. value (in ppm) is 13 times 0.95, or 12.4 ppm.

If less than 20 ml were required, the solution is stronger than 0.025M and the correction value is more than 1. For example, if 17 ml were used, divide 20 by 17 to give 1.18, the correction value. If 8 ml of the sodium thiosulfate are used in a D.O. titration, the corrected D.O. value in ppm is 8 times 1.18, or 9.4 ppm.

Procedure

a)   Completely fill a 250–300 ml sample bottle (preferably a B.O.D. bottle) with the water to be tested. Be sure no air enters the water during this process.

b)   Add 2 ml of manganese(II) sulfate solution and 2 ml of alkaline-iodide solution, *beneath the liquid surface.* Use a separate pipet for each solution. Caution: Do not mix the water, or contamination with the atmosphere will result. Be sure that the excess solution overflows onto a surface that cannot be harmed by it.

c)   Replace the stopper, making sure that no air bubbles are trapped beneath it. Shake the mixture vigorously and then allow the precipitate to settle to the lower half of the bottle.

d)   Add 2 ml of concentrated sulfuric acid above the water level; replace the stopper carefully. Shake. At this point titration can be delayed several hours without affecting the result.

e)   Transfer 200 ml to a 500 ml Erlenmeyer flask using a volumetric pipet.

f)   Titrate with $0.025M$ sodium thiosulfate solution. One or 2 drops of the starch solution should be added when the color of the solution has become light yellow (after the addition of some of the sodium thiosulfate). Continue titrating until the blue color is gone.

g)   The D.O. in ppm is the number of ml of $0.025M$ sodium thiosulfate solution used to reach the end-point, subject to correction as noted above.

Notes

a)   All steps before the final titration must be done in the field immediately upon collection of the water sample. The final titration can be done a few hours later.

b)   Caution must be used in handling these chemicals.

c)   All of the required solutions can be purchased, pre-standardized, from most chemical supply companies.

d)   If you have studied chemistry, you should look up the reactions that are involved in this D.O. determination.

6.8   FREE CARBON DIOXIDE

The carbon dioxide content of waters is important from an economic viewpoint, since this gas contributes to several forms of corrosion. Biologically, a concentration of carbon dioxide greater than 25 ppm can be lethal to aquatic animals. Further, a high carbon dioxide concentration is usually accompanied by a low D.O. concentration.

     Test kits of the Hach and LaMotte type are available for measuring the carbon dioxide content of water. The following procedures can also be used.

Materials

a)   several 100 ml, short form (33 × 200 mm) Nessler tubes

b)   droppers

c)   buret

d)   phenolphthalein indicator solution

e)   0.0227M sodium carbonate solution

Procedure

a)   Collect 100 ml of the sample water in a Nessler tube. Avoid agitation and contact with the air. Proceed as quickly as possible with the following steps in this procedure.

b)   Add 10 drops of the phenolphthalein indicator. If a pink color forms, no carbon dioxide is present.

c)   If the sample stays colorless, titrate with the 0.0227M sodium carbonate solution until a faint but permanent pink color forms. Do not agitate the sample during the titration, but rotate the tube gently in order to mix. This avoids contamination from atmospheric carbon dioxide.

d)   The free carbon dioxide present in the water, in ppm, is the number of ml of sodium carbonate solution used multiplied by 10.

6.9   pH

pH can be measured by several methods. One easy and fairly reliable method uses a pH kit with a universal indicator and a color comparator (Fig. 6-5). The accuracy of such a kit is usually limited to ± 0.5 pH units. Just as convenient, reliable, and accurate are the Fisher Alkacid test papers (Fig. 6-6). If higher

Fig. 6-5 (Left)
A pH kit that uses a universal indicator and color comparator.

Fig. 6-6 (Right)
Fisher Alkacid test papers.

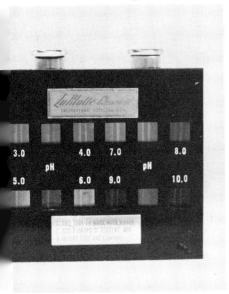

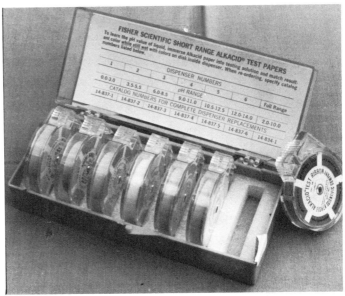

accuracy is required, a portable pH meter can be used. Such devices are expensive though, and often difficult to maintain. Accuracy almost as high as with a pH meter can be obtained using the method outlined in Section 5.3 of Andrews et al., *A Guide to the Study of Freshwater Ecology* (Prentice-Hall, 1972).

6.10   ALKALINITY

Tests for alkalinity are easily done with small kits (Hach and LaMotte). However, if you wish to do your own chemistry, the test is described below.

Materials
a)   250 ml Erlenmeyer flask
b)   droppers
c)   buret
d)   100 ml volumetric pipet
e)   phenolphthalein indicator solution
f)   methyl orange indicator solution
g) $0.01M$ sulfuric acid

Procedure
a)   Add 5 drops of the phenolphthalein indicator to a 100 ml sample of the water to be tested.
b)   If the solution becomes colored, proceed with step c); if not, go directly on to step d), skipping c).
c)   If the solution became colored in step b), titrate with $0.01M$ sulfuric acid until the color disappears. Record the number of ml of acid used. To find the phenolphthalein alkalinity, in parts per million (ppm) as $CaCO_3$, multiply the number of ml of acid used by 10. (The phenolphthalein alkalinity measures the amount of hydroxide and half the amount of carbonate in the water.) Continue with step d).
d)   Add 5 drops of methyl orange indicator to the solution from step b) or c). If the solution becomes yellow, titrate with $0.01M$ sulfuric acid until a pinkish color appears and persists. Record the ml of acid used.
e)   The methyl orange or total alkalinity is the sum of the ml of acid required for both of the titrations, multiplied by

10. This gives the total alkalinity in ppm as $CaCO_3$. Total alkalinity includes hydroxide, carbonate, and bicarbonate.

Note: All titrations should be done over a white surface so that color changes can be spotted quickly.

6.11   TOTAL HARDNESS (T.H.)

Hardness can be measured best with any one of the kits of the Hach or LaMotte variety. You can even buy kits for this purpose from pet shops.

The soap method for determining hardness is described below. No attempt has been made to include ways of correcting for the lather factor, since in most cases an error of $\pm$ 10 ppm is acceptable. With several commercially available water test kits, errors can be as high as $\pm$ 16 ppm.

Materials

a)   250 ml glass-stoppered bottle

b)   50 ml pipet

c)   droppers

d)   500 ml and 1 liter volumetric flasks

e)   stock soap solution: shake 100 gm of pure castile soap powder into 1 liter of 80% ethyl alcohol. Cover and let stand for 1 to 3 days, then decant the upper layer. Discard the soap left in the bottom of the bottle.

f)   standard soap solution: dilute a small portion of the stock soap solution with 80% ethyl alcohol until 1 ml is equivalent in hardness to 1 ml of standard calcium chloride solution.

g)   standard calcium chloride solution: dissolve 0.5 gm of anhydrous calcium chloride in a few ml of dilute hydrochloric acid. Add 200 ml of carbon dioxide-free distilled water. Neutralize the solution with ammonium hydroxide until it is just alkaline (use litmus as an indicator). Add further distilled water until the volume of the solution is 500 ml. Store in a glass container. One ml of this solution is equivalent in hardness to 1 mg of calcium carbonate, which, in turn, is equivalent in hardness units to 1 ml of standard soap solution. Thus it is necessary for you to test the soap solution and adjust its concentration until 1 ml of the solution just forms a permanent froth when it is shaken with 1 ml of the standard calcium chloride solution.

Procedure

a) Place a 50 ml water sample into a 250 ml bottle.

b) Add standard soap solution in 0.2 ml drops. Shake the bottle after each drop.

c) When a lather is first seen, let the bottle stand for 5 minutes. If the lather remains, the end-point is reached. If the lather disappears, continue to add soap solution until the lather does remain.

d) The number of ml of soap solution used, multiplied by 20, gives the hardness (in terms of calcium carbonate) in ppm.

6.12 ACIDITY

Acidity can be measured best with a kit (Hach or LaMotte). If you do not have a kit, you can use the method below.

Materials

a) 50 ml pipet

b) 250 ml Erlenmeyer flask

c) buret

d) phenolphthalein indicator solution

e) bromcresol green–methyl red indicator solution

f) $0.020M$ sodium hydroxide solution

Procedure

a) Transfer a 50 ml sample of the water to be tested to a 250 ml Erlenmeyer flask. Avoid agitation to minimize the loss of carbon dioxide gas.

b) Add a few drops of phenolphthalein indicator solution.

c) Titrate with $0.020M$ sodium hydroxide solution until a permanent pink color appears. The total acidity as ppm of $CaCO_3$ is equal to the number of ml of sodium hydroxide solution multiplied by 20.

d) Repeat step a).

e) Add a few drops of bromcresol green–methyl red indicator solution.

f) Titrate with 0.020M sodium hydroxide solution until the color changes from red to a light pinkish-grey with a blue cast. The free acidity as ppm of $CaCO_3$ is equal to the number of ml of sodium hydroxide solution multiplied by 20.

6.13   NITROGEN AND PHOSPHORUS

Hach or LaMotte kits should be used to test for nitrogen (ammonia, nitrite, and nitrate) and for phosphorus (phosphate). All other acceptable methods are somewhat complex and require expensive colorimeters. However, if your school has a suitable colorimeter, you may want to try some of these methods. They are outlined in *Standard Methods for the Examination of Water and Waste Water*, published by the American Public Health Association.

6.14   TOTAL SUSPENDED SOLIDS (T.S.S.)

Materials

a) fine filter paper

b) analytical balance

c) 1 liter bottle

d) funnel

Procedure

a) Weigh a filter paper.

b) Filter a 1-liter sample of water through the weighed filter paper.

c) Allow the filter paper to dry completely.

d) Reweigh the filter paper. The change in weight is the weight of the total suspended solids (T.S.S.) in 1 liter of water. T.S.S. values are commonly expressed in ppm (mg per liter).

6.15   TOTAL DISSOLVED SOLIDS (T.D.S.)

The experiment for testing the T.D.S. is rather simple, but it is exactingly quantitative. Accurate tests cannot be easily done except in a laboratory equipped for them. The method below, however, will give you reasonably accurate results.

Materials

a)   250 ml beakers
b)   dust-proof chamber (if available)
c)   analytical balance
d)   electric hot plates
e)   volumetric pipet

Procedure

a)   Weigh to the nearest 0.0001 gm, if possible, 3 clean, dry 250 ml beakers.

b)   Place 100 ml of the filtrate from the T.S.S. experiment (Section 6.14) in each of these beakers.

c)   Slowly and carefully evaporate to dryness using electric hot plates and a dust-proof chamber. Do not let the beakers get too hot or some of the dissolved solids may be vaporized or decomposed.

d)   Make the appropriate calculations for each of the 3 samples to determine the weight in grams of the solids dissolved in 100 ml of water. Average the 3 values that you obtain.

e)   Convert your answer to ppm by multiplying the value obtained in d) by 10,000.

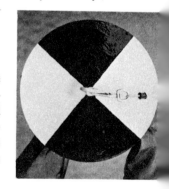

Fig. 6-7
A Secchi disc, used for determining the transparency of a body of water.

6.16   TRANSPARENCY

An indication of the amount of suspended matter in water can be obtained with a Secchi disc. This is a metal disc, 20 cm in diameter and divided into four quarters, two of which are white and two black (Fig. 6-7). Several kinds of information can be gathered with the disc: a rough measure of the suspended matter; the depth of reflected light penetration; and a rough estimate of the extent of the littoral zone.

To obtain a Secchi disc reading, lower the disc into the water, in the shade, until it just disappears. Take a depth measurement at this point. Then raise the disc until it just reappears. Take another depth measurement. The two measurements are averaged. This process is repeated three times and the overall average is considered to be the proper Secchi disc reading. The surface conditions and the color of the water should be noted when performing this test.

If the Secchi disc reading is low (for example 5 feet), the water contains much suspended matter. If, on the other hand, it is very high (for example 30 feet) the water is quite clear and hence relatively free of suspended matter.

It has been suggested that the Secchi disc reading corresponds closely to the depth limit of the littoral zone, the area of rooted plant growth. Thus, if the Secchi disc reading is 15 feet, the 15 foot contour of the lake is roughly the boundary of the littoral zone.

The Secchi disc reading, however, can vary from place to place, from hour to hour, and from day to day, depending on above-water conditions. When you are trying to determine the location of the littoral zone, test repeatedly over a period of a few months, in order to get the best approximation of the zone boundary.

6.17 COLOR

The color of water gives an indication of the amount of suspended and dissolved matter present in it.

The best way to find the color of water is with a color comparator. One form of the standard Forel-Ule Scale is shown in Figure 6-8. The water color is most easily determined in conjunction with the Secchi disc. Lower the disc until it lies about 3

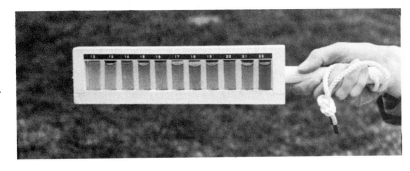

Fig. 6-8
A Forel-Ule Scale for determining water color.

feet below the surface. The number of the vial that blends most closely with the water color against the Secchi disc is the color number. The whiteness of the disc provides the background to which the color is referred. The vials should be shaded from direct sunlight when the determination is made, since reflections could distort the readings.

If the color comparator and Secchi disc are not available, you can often make a reasonable color determination by holding a test tube full of the sample against a white background. Adjectives such as light, medium, and dark can be used to describe the color. It is best for the sample to stand for a few minutes to allow solids to settle. An inexpensive Hach color test kit is also available.

6.18   TEMPERATURE

Temperature plays an important role in determining the species of organisms which can live in a particular body of water. As you will recall, it also affects D.O. levels.

Air temperature determines, to a large extent, the surface water temperature. An ordinary thermometer can be used to take the air temperature, but the reading must be taken in the shade. (Hold your hand between the sun and the bulb of the thermometer.)

Generally, for water temperature a *series* must be taken. In deep lakes this aids in locating the thermocline; in shallower water, it shows whether there is a thermocline, or even the start of one.

For waters less than 2 meters in depth, readings at the surface and at the bottom are usually sufficient. For waters more than 2 meters in depth, a reading should be taken at 1 or 2 meter intervals. For critical areas, however (at the top and bottom of the thermocline, for example), readings should be taken at 0.5 meter intervals.

The best method of obtaining a temperature series is with a *thermistor*. This is a small, battery-operated unit which usually has readout scales in both °F and °C. A maximum–minimum thermometer can also be used, but this requires hauling the unit back to the surface each time to record the temperature and to reset the indicator markers. The least accurate method is to use a collecting bottle to obtain a water sample from the desired depth. Quickly raise it and take the temperature before it has a chance to change too much.

An indoor–outdoor thermometer of the type shown in Figure 6-9 is of real value in determining simultaneously the air

Fig. 6-9
An indoor–outdoor thermometer, used in this manner, will give readings of the air and water temperature simultaneously.

and water temperatures. Up to depths of 2 meters or so, this instrument can also be used to determine quickly and accurately water temperatures at different depths.

6.19 VELOCITY OF FLOW

Section 2.8 pointed out the importance of considering velocity of flow during a study of a stream or river. Velocity of flow influences many other factors such as D.O. concentration, carbon dioxide concentration, and temperature. A pollution study would be incomplete without a determination of this physical factor.

Materials

a)   stopwatch
b)   known length of string
c)   buoyant object such as an orange or a styrofoam ball

Procedure

Stand near the center of the stream. Set the stop watch and place the object on the end of the string into the water. At the moment the object enters the water, start the watch. As soon as the string becomes tight, stop the watch. During this procedure, the hand holding the string should be as near as possible to the surface of the water.

You have now recorded the *time* and the *distance* (length of the string). If the string is measured in meters, you can very easily compute the velocity in *meters per second*. Repeat this procedure 3 or 4 times and average the results.

Two additional methods for determining velocity of flow are found in Section 5.13 of Andrews et al., *A Guide to the Study of Freshwater Ecology* (Prentice-Hall, 1972).

6.20 VOLUME OF FLOW

The volume of flow is important in pollution studies since it determines, to a large extent, the ability of a stream to handle pollutants that are added to it.

Materials

a)   stopwatch

b)   float

c)   tape measure

Procedure

Determine the following values:

t: the time in seconds required for the float to travel a measured section of a stream;

l: the length in meters of the stream section;

w: the average width in meters of the stream section;

d: the average depth in meters of the stream section.

To compute the rate or volume of flow in cubic meters per second, use the following formula:

$$r = \frac{wdal}{t}$$

where a is a constant. The value of a is 0.8 if the stream bed is composed of rubble or gravel, and 0.9 if the stream bed is quite smooth (sand, mud, silt, or bedrock).

If you have already measured the stream velocity in meters per second as outlined in Section 6.19, use the formula

$$r = wdav$$

where v is the velocity of the stream.

6.21   CROSS-SECTIONAL PROFILE OF A STREAM

A pollution study of a stream should include a cross-sectional profile of the stream. The profile may explain certain variations in temperature, velocity, and types of life across the width of the stream.

Materials

a)   string

b)   meter stick

Procedure

Suspend a string across the width of the stream, its ends tied securely at each side. At suitable intervals along the string, measure depth of the water, type of bottom material, and amount of vegetation. (For a stream 1 m wide, a suitable interval is 5 cm; for a stream 5 m wide, 0.5 m.) In recording bottom type, you can use the system below or you can construct one of your own.

Bedrock	—————	Sand	xxxxxxxx
Mud or Silt	··············	Gravel	oooooooo
Rubble	********		

When you have recorded your data, use a suitable scale to reconstruct the stream profile on paper.

6.22   BIOCHEMICAL OXYGEN DEMAND

The apparatus required for a good B.O.D. determination is quite expensive. A substitute method is given here for determining the *relative stability* of water samples. Relative stability values are not as useful as actual B.O.D. values. They do, however, give an indication of the oxygen demand of the water. This, in turn, indicates the extent to which the water is contaminated with biologically oxidizable matter.

Materials

a)   methylene blue solution: 0.10 gm of methylene blue dissolved in 100 ml of distilled water.

b)   250 ml bottle with ground-glass stopper (preferably a B.O.D. bottle)

c)   1 ml pipet

Procedure

a)   Fill the B.O.D. bottle with the water to be tested. Do not allow any air to enter the bottle.

b)   Add about 1 ml of methylene blue solution. The tip of the pipet should be in the water to prevent the methylene blue from absorbing oxygen from the air.

c)   Replace the stopper. Place the bottle in a warm, dark place. The temperature should be as close to 20°C as possible.

d)   Check the bottle every 12–24 hours. Watch for the disappearance of the blue color.

e)   Use Table 17 to calculate the relative stability of the water sample.

TABLE 17

Number of days for blue color to disappear	Relative stability (percent)
0.5	11
1.0	21
2.0	37
3.0	50
4.0	60
5.0	67
6.0	75
7.0	80
8.0	84
9.0	87
10.0	90
11.0	93
12.0	95

The methylene blue will remain blue so long as oxygen is present. When the oxygen in the water is used up (largely by decomposer organisms acting on organic matter), the blue color disappears. The table tells you the percentage of the total biochemical oxygen demand that has been satisfied. If, for example, 8 days are required for the blue color to disappear, the sample had sufficient oxygen to satisfy 84% of the total biochemical oxygen demand. This sample did not contain a great deal of oxidizable matter. Otherwise, its oxygen supply would have been used up much sooner.

6.23   **CHEMICAL OXYGEN DEMAND (C.O.D.)**

The Hach Chemical Company supplies the materials for an inexpensive but effective determination of C.O.D. The procedure is described in Hach Catalog Number 10, "Water and Wastewater Analysis Procedures." Study this procedure before you decide to make C.O.D. determinations a part of your pollution studies.

6.24 DETERGENTS

The Hach Chemical Company has a detergent test kit that will detect 0–1 ppm of both L.A.S. (biodegradable) and A.B.S. (non-biodegradable) detergents. This kit (Model DE-2) contains its own color comparator and is quite easy to use. Since sewage effluent normally contains more than 1 ppm of detergent, it is necessary to do a 1 to 10 dilution when analyzing sewage effluent for detergents.

6.25 EQUIPMENT AND TECHNIQUES FOR BIOLOGICAL SAMPLING

Sections 6.26–6.29 outline how to conduct field trips to polluted bodies of water. The physical and chemical tests for such studies were described in earlier sections of this unit. Be sure you can perform these tests before going on a field trip. The success of these studies also depends on your ability to sample effectively the living organisms in the water. Listed under *Materials* in these sections are some important pieces of equipment that you can use to capture living organisms. The techniques for using these collecting devices are outlined in Andrews et al., *A Guide to the Study of Freshwater Ecology* (Prentice-Hall, 1972). Consult this reference and summarize the specific techniques that you require for your studies.

6.26 FIELD TRIP TO A POLLUTED STREAM

Place

Find a situation similar to that shown in Figure 6-10. Such a situation is easy to find since towns and cities are commonly located on waterways. The river or stream flows past or through the municipality and receives street runoff, effluent from industries, and wastes from the sewage treatment plant. The obvious study here is a comparison of the water quality above and below the municipality.

For safety reasons avoid study sites where the stream is wider than 20 feet and deeper than 3 feet. Remember, too, that many waterways become dangerously flooded after heavy rains and during the spring.

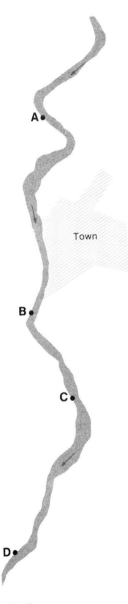

Fig. 6-10
A study of the water quality of a stream.

Materials

a) collecting jars, vials, and buckets
b) preservative (25% formalin for fish, 70% ethanol for invertebrates)
c) paper labels and marking pens
d) notebooks
e) topographical map of the area
f) identification guide books
g) rubber boots and chest waders
h) rubber gloves
i) water testing kits (D.O., alkalinity, etc.)
j) thermometers
k) water velocity equipment: stopwatch, measuring tape, string
l) profile equipment: string, rulers
m) sieves
n) plankton net
o) seine net
p) dip nets
q) lifters
r) sorting trays and forceps
s) Surber sampler and Ekman dredge, if available
t) microscopes: compound, dissecting
u) microscope slides, eyedroppers for plankton analysis

Procedure

a) Select study sites as follows:
   A: upstream from the town but as close to the town as possible.
   B: downstream from the town, preferably within 100 yards or so of the town's major sewage outlet.
   C: 1 or 2 miles downstream from the town (to determine the degree of self-purification that has occurred).
   D: 2 or 3 miles further downstream from C, provided no sources of pollution intervene (to determine if further self-purification has occurred).
b) At each site study the physical characteristics of the stream: cross-sectional profile, velocity of flow, volume of flow, temperature, color, transparency.

c) At each site do a chemical study of the water: D.O., free carbon dioxide, pH, alkalinity, T.H., acidity, ammonia, nitrate, phosphate, T.S.S., T.D.S., detergents, B.O.D., and C.O.D. Some of these studies can, of course, be completed back in the laboratory.

d) At each site sample the biological organisms to determine the species present and their relative abundance: total coliforms, *E. coli*, bottom fauna, algae, zooplankton, and fish.

Notes

a) Keep in mind that some activities in the stream will disrupt other planned activities. The chemical tests should, in general, be done before the water and bottom silts are disturbed. Plankton samples should be taken early. The biological sampling could follow. The invertebrate collections should be completed and properly preserved, stored, and labeled before seining and other fish collecting is even attempted. The physical factors could be analyzed last.

b) Mapping or sketching the study area is useful in defining the extent of each study site and for showing the location of specific studies such as the determination of bottom conditions, depth, position of profiles, and water velocities.

c) This study requires the cooperation of several groups of people. Be sure that the tasks of each group are clearly defined before the outing. For example, one group of 3 people may be assigned the task of performing D.O., T.H., and alkalinity tests.

d) The field trip can be shortened by eliminating site D and, if necessary, site C. It is difficult for one class of 30 students to study more than 2 sites in one day. Two classes could combine their efforts and study all 4 sites in one day.

e) Make a note of the date, time of day, and weather conditions.

Follow-up

Back in the laboratory you should attempt to analyze the observed differences in physical, chemical, and biological conditions at the 4 study sites. Keep your mind open; assess the data scientifically. Don't blame pollution for all of the biological changes that you observe unless you have conclusive evidence. For example, stonefly nymphs may live upstream from the town but not downstream solely because of a difference in natural bottom conditions.

If conditions of the stream need improvement, either above or below the town, a management plan in the form of a map could be drawn to show the position and the type of improvement that you think would be helpful. You may wish to submit this plan and a report of your findings to the town council and to any government agency that is responsible for the maintenance of water quality in that particular waterway.

6.27   FURTHER STUDIES OF POLLUTED STREAMS

The Effect of a Weir on Water Quality

A weir is often constructed across a stream or river in an attempt to improve the water quality (Fig. 2-8). A class of 30 students can, in 3 or 4 hours, determine just how effective the weir is. To do so, establish 2 study sites, one just upstream from the weir and the other about 50–100 yards downstream from the weir. At each site make a thorough study of physical, chemical, and biological factors as recommended in Section 6.26. A comparison of the data from the 2 sites should indicate clearly the effectiveness of the weir.

Before you begin this study you should make a list of those factors that you expect to be affected by the weir.

The Effect of a Heavy Rain on Water Quality

A heavy rain may improve the water quality of a stream or river by diluting and washing away pollutants. It may also lower the water quality by washing into the waterway such things as fertilizers, barnyard runoff, and suspended solids. Further, during heavy rainstorms the sewage systems of many municipalities become overtaxed, and untreated sewage flows into the waterway. The object of this study is to determine whether a heavy rain improves or lowers the water quality of a particular river or stream. If your data indicate that the water quality is lowered, they will also suggest what the cause might be.

The first thing that you must do in this study is to establish a set of norms for the river or stream. Select a site that can be conveniently visited several times. Pay four or five visits to this site, all at the same time of day. Each visit must be made on a day when there has been no appreciable rainfall for at least 48 hours prior to the visit. On each visit, carry out a thorough study of the physical, chemical, and biological aspects of the site. The combined data from the four or five visits give you an idea of the "normal" state of the waterway.

The final visit is made a few hours after a heavy rain. Study the same physical, chemical, and biological factors on this visit as you did on the others. Account for any observed changes. Volume of flow, T.S.S., and T.D.S. determinations should be given particular attention in this study.

Seasonal Variations in Water Quality

Precipitation and temperature are two physical factors that have a marked effect on water quality. In most parts of North America both of these factors vary considerably over the course of the year. It is of interest, therefore, to study the same site in a river or stream at several different times of year: spring, early summer, late summer or early fall, late fall, and winter.

Each study should be conducted on a day that appears to be normal for that time of year. The two days prior to the study should also have been normal. If time permits, a more precise way of conducting this investigation is to establish a norm for each season using the procedure outlined above. Since no one class is likely to be allowed enough outings to complete this study, it is suggested that this be a school project. After the data have been collected, a meeting of the participating classes should be held to discuss and interpret the seasonal variations.

6.28 FIELD TRIP TO A POLLUTED SMALL LAKE

Place

Select a relatively small lake that receives effluent from a municipality or from industries. About 100 acres or less is an ideal size. A larger lake is too difficult to sample effectively. Find out as much as you can about the past history of the lake. Look into its geological origin and the geographical features of the surrounding countryside. Consult a geography or geology teacher in your school for reference materials such as topographical maps and soil maps. Consult the government agency that is responsible for maintaining the quality of the water in the lake. This agency has probably performed for many years the tests that you are going to perform. This study of the past history of the lake will help you to decide whether your findings are the result of pollution or simply a reflection of natural eutrophication. The major objective of this field trip is to determine the degree of eutrophication of the lake and the possible causes of the eutrophication.

Materials

a)   collecting jars, vials, and buckets
b)   preservative (25% formalin for fish, 70% ethanol for invertebrates)
c)   paper labels and marking pens
d)   notebooks
e)   topographical map
f)   identification guide books
g)   rubber boots and chest waders
h)   rubber gloves
i)   life preservers
j)   boat (2- or 4-man rubber dinghies are safe and easy to work from)
k)   water testing kits (D.O., alkalinity, etc.)
l)   thermometers
m)   water velocity equipment for inlets and outlets
n)   measuring tape
o)   water sampling bottle (Kemmerer or homemade)
p)   Secchi disc
q)   Ekman dredge
r)   sorting trays and forceps
s)   plankton net
t)   sieves
u)   seine net (6 foot, although for a larger area, nets up to 25 feet can be used)
v)   dip nets
w)   lifters
x)   microscopes: compound and dissecting
y)   microscope slides and eyedroppers for plankton analysis

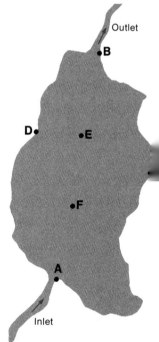

Procedure

a)   If possible, select study sites as shown in Figure 6-11.
     A: at the inlet
     B: at the outlet

C and D: sites remote from the inlet
E and F: mobile sites (requiring boats)

b)   The appropriate physical tests should be performed at each site: inflow determination at A; outflow determination at B; transparency determination by teams at E and F; color and temperature at all sites.
Note: The mobile teams (E and F) should perform the last 2 tests on water from several depths.

c)   At each site do a chemical study of the water: D.O., free carbon dioxide, pH, alkalinity, T.H., acidity, ammonia, nitrate, phosphate, T.S.S., T.D.S., detergents, B.O.D., and C.O.D. Some of these studies can, of course, be completed back in the laboratory. The mobile teams (E and F) should perform these tests on water from several depths.

d)   At each site, sample the organisms to determine the species present and their relative abundance: total coliforms, *E. coli*, bottom fauna, algae, zooplankton, and fish.

Notes

a)   Map or sketch the study area. Mark on the map the extent of each study area and unique geographical features around the lake.

b)   Record the date, time of year, and weather conditions. In general, studies should not be undertaken within 24 hours of a heavy rainfall, unless you wish to compare the results with normal conditions.

c)   The data collected by teams E and F should indicate whether overturn (spring or fall) has occurred. Why is it important to know this? Early autumn is an ideal time to perform a lake study. Why?

Follow-up

Back in the laboratory you should attempt to analyze the observed differences in physical, chemical, and biological conditions at the chosen locations and depths. Don't draw conclusions without considering the past history of the lake. If conditions require improvement, draw up a list of suggestions and prepare a management plan. You may wish to submit your findings, suggestions, and management plan to the government agency that is responsible for the maintenance of water quality in that lake. Local industries and municipal governments could also be made aware of your work.

6.29 POLLUTION STUDIES ALONG A LARGE LAKE

If you live near a large lake (for example, one of the Great Lakes), you may want to do a pollution study of that body of water. For obvious reasons you cannot perform the kind of study outlined in Section 6.28. You can, however, carry out an interesting and meaningful study with the guidelines below.

Select your first study site as close as possible to one of the following:

1)	the mouth of a river that carries pollutants into the lake;
2)	a sewage treatment plant that dumps its effluent into the lake;
3)	an industry that dumps its effluent into the lake or one that uses the lake water as a coolant.

Select a second study site on a beach that could be affected by the pollution coming from your first site. A public swimming beach within a mile or so of the pollution source is ideal. Consider the prevailing water currents when selecting this site.

Select a third study site on a beach that is probably not affected directly by the pollution coming from your first site. Again, a knowledge of the prevailing water currents is necessary when selecting this site. Make sure that this beach is not affected by another source of pollution further up the beach.

At each site do all of the physical, chemical, and biological tests that can be safely performed. Carry out a follow-up as outlined in Section 6.28. If you find a public beach that is closed to swimming because of pollution, try to find out from your municipal authorities why it was closed, how long it has been closed, and what plans are being made to restore the water quality.

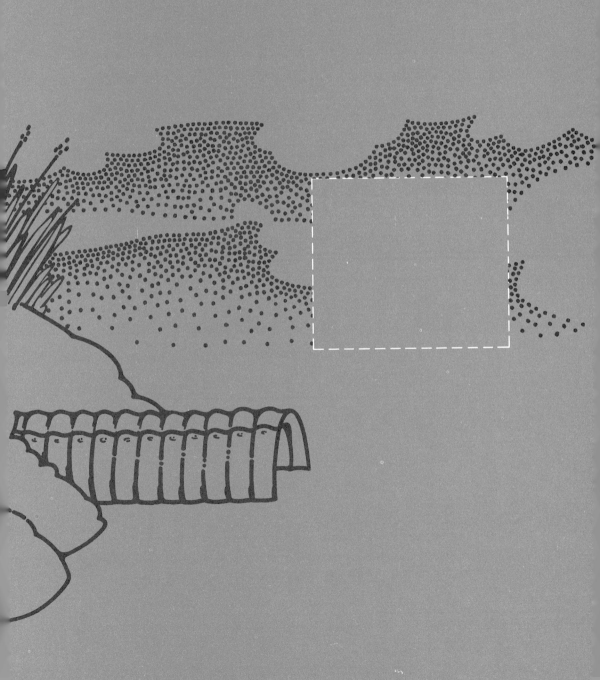

Field and Laboratory Studies in Air Pollution

7

7.1 DUSTFALL PER UNIT AREA

How many pounds of solid particles fall on your front lawn or school yard during an average month? Does your rural neighbor or the nearest factory receive the same quantity? By selecting a number of test sites in different areas, experimental data can be compared over a wide range of the region in which you live.

A. COLLECTION OF SAMPLES

Materials
a) gallon jar with screw lid (The mouth should have a minimum diameter of 4 in.)
b) wooden base to support the jar
c) wire with suitable attachments to secure the jar to the base
d) a 20-mesh screen to cover the jar mouth. This prevents large objects from entering the jar.
e) 1 liter of distilled water

f)   ammonium chloride to inhibit the growth of algae (1.5 ppm by weight of distilled water)

g)   isopropyl alcohol (100 ml) to prevent freezing during low temperature periods

Procedure

For each sampling site:

a)   Add 1 liter of distilled water to a clean, dry gallon jar.

b)   Add 2.0 mg of ammonium chloride (2 ml of a solution formed by dissolving 0.1 gm of ammonium chloride in 100 ml of distilled water).
Note: The amount of ammonium chloride added should be recorded for future calculations.

c)   If freezing temperatures are expected, add 100 ml of isopropyl alcohol to the distilled water.

d)   Mark the liquid level on the outside of the jar using a waterproof marker. Place the lid on the jar.

e)   Select a wide variety of sampling sites. For example, test an industrial area, a residential area, the school site, the nearest farm or conservation area, or the nearest summer resort area. A school in the city and one in the country could work together on this project. Or, several schools in a community could work together to pin-point the dirtiest areas in the community. Each site must be freely exposed to gravitational settling from the air. Try to avoid interference from surface dirt on surrounding buildings or other high objects. The apparatus should also be inaccessible to the general public.

f)   At each site, secure the wooden base on a flat surface at least 6 feet above ground level to avoid surface dust. Remove the jar lid and cover the mouth with the 20-mesh screen. Secure the screen and jar to the base using the support wires (Fig. 7-1).

g)   Expose each jar for 30 days. Further additions of distilled water should be made regularly to maintain the original liquid level.

h)   When the test period is complete, rinse each wire screen with distilled water to wash into the jar accumulations which settle on the wire. Then remove the screen, replace the lid on the jar, and transport the apparatus back to the laboratory.

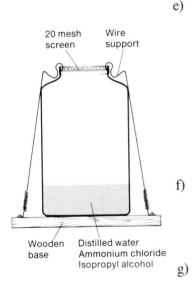

20 mesh screen   Wire support

Wooden base   Distilled water
Ammonium chloride
Isopropyl alcohol

Fig. 7-1
The sample collection jar should be secured to a support base.

B. DETERMINATION OF DUSTFALL PER UNIT AREA

Materials

a) 250 ml beaker
b) 2 liter beaker
c) distilled water
d) rubber policeman
e) 0.025% sodium lauryl sulfate solution
f) drying oven
g) hot plate
h) analytical balance
i) desiccator

Procedure

a) Pour the liquid from the jar into a clean 2 liter beaker.

b) Use the rubber policeman and small portions of hot distilled water to loosen solids clinging to the sides of the jar. Pour this liquid into the 2 liter beaker. If some solids still adhere to the sides of the jar, loosen them with 0.025% sodium lauryl sulfate solution (a detergent). Use no more than 15 or 20 ml. Record how much you use.

c) Clean a 250 ml beaker. Dry it for at least 1 hour in an oven at 105 °C. Let it cool in a desiccator and weigh it to the nearest 0.1 mg.

d) In small portions transfer all of the liquid in the 2 liter beaker to the 250 ml beaker, evaporating most of the water after each transfer. The evaporation can be done with either a hot plate or an oven. Do not boil the liquid; splattering will cause the loss of some of the sample. Also, excessive temperatures may vaporize some of the solids.

e) Evaporate the residue in the 250 ml beaker to dryness in an oven at 105 °C for at least 1 hour.

f) Let the beaker and contents cool in a desiccator. Weigh to the nearest 0.1 mg.

Calculations

a) Determine the weight of the solids in the 250 ml beaker by subtracting the weight of the beaker from the weight of the beaker plus its contents.

b)   If ammonium chloride or sodium lauryl sulfate were used, subtract their *weights* from the weight obtained in a). This final figure is the total dustfall that entered your container during the 30-day period.

c)   Calculate the area of exposure of the jar in cm². (Measure the inside diameter, d, of the mouth of the jar to the nearest 0.1 cm and substitute in $A = 0.785\ d^2$.)

d)   Calculate the dustfall in mg per cm² of surface area per 30 days.

e)   You may wish to convert your answer to pounds per square foot or tons per square mile to more easily visualize the quantities involved. Many air pollution authorities use these units. Try the conversions. If you have any problems, consult your teacher.

Discussion

Compare the results obtained from the various sampling sites. Account for any differences. If dustfall is recorded by air pollution authorities in your area, compare your results with theirs.

7.2 PARTICULATE MATTER PER UNIT VOLUME OF AIR

The following outline describes the use of Millipore equipment to determine the weight of particulate matter in a unit volume of air.* You should perform the investigation at several sites. You should also test one or two sites at the same time each day for several weeks. Be sure to record wind direction, wind velocity, and other prevailing weather conditions at the time of each sampling.

Materials

a)   Millipore Sterifil filter holder

b)   Type AA (0.8 micrometer pore size) Millipore filter

c)   Millipore aerosol adapter

d)   set of Millipore limiting orifices

e)   analytical balance

f)   forceps

*Provided through the courtesy of Millipore Corporation, Bedford, Mass. Copyright® Millipore Corporation (1963).

g) petri dish

h) vacuum source such as a vacuum pump or vacuum cleaner. For field work, a vacuum cleaner that plugs into the cigarette lighter of a car is ideal.

Procedure

a) Weigh a filter to the nearest 0.1 mg.

b) Load the weighed filter into the filter holder.

c) Attach the hose connection of the filter holder to an aerosol adapter containing a suitable limiting orifice. Connect the aerosol adapter to the vacuum source (Fig. 7-2).

d) Draw air through the filter using Table 18 as a guide. These are minimum values. Larger samples are often required.

e) Remove the filter carefully with forceps. Weigh it to the nearest 0.1 mg.

f) Calculate the change in weight of the filter. This figure represents the weight of particulate matter present in the air that passed through the filter. Convert this to mg per liter.

g) Place the filter in a petri dish and seal it. Save it for Section 7.4.

Discussion

Account for the variation in the values obtained at the different sites. Explain the variations that occur at any one site over a long period of time.

Some of the weight change of the filter paper is due to moisture absorption and other environmental factors. These factors can be accounted for with the control filter method outlined in *Microchemical and Instrumental Analysis*, Millipore Corp.,

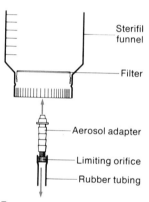

Fig. 7-2
Apparatus for particulate sample collection.

— Sterifil funnel

— Filter

— Aerosol adapter

— Limiting orifice

— Rubber tubing

To vacuum system

TABLE 18

	Rate	Time
Room air	10 liters/min	for 28 min
Country air	10 liters/min	for 14 min
City air	1 liter/min	for 28 min
Factory air	0.5 liter/min	for 28 min

1963. This same book describes the use of matched weight filters to speed up and increase the accuracy of testing, particularly in the field.

7.3 MICROSCOPIC EXAMINATION OF PARTICULATE MATTER

Particulate material can be partially analyzed under the microscope. The origin and composition of a particle can often be deduced from its size, shape, color, and texture. Metal shavings, for example, are readily distinguishable from sand grains or dust. Wool and other animal hairs have typical, scaly patterned surfaces. A guide to the identification of single fibers is available from Millipore. Collect particulate samples at a variety of sampling sites. Construction sites, underground parking lots, the airport, factory areas, and congested streets are a few examples.

A. THE STEREO MICROSCOPE

The three-dimensional presentation of the stereo microscope is often a distinct advantage in the identification of particles. However, since this microscope is limited to low powers, it is only useful for relatively large particles.

Procedure

a) Expose a glass slide, coated with stopcock grease, petroleum jelly, or a similar transparent adhesive, to dustfall at selected sampling sites until a suitable sample has settled. (The slide could be attached to the base of the apparatus used for dustfall determination in Section 7.1.)

b) Examine the sample carefully to determine the type of particles present.

B. EXAMINATION USING OBLIQUE INCIDENT LIGHT

Large particulate material (greater than 10 microns) can be studied best using this method. Surface detail, color, and texture should be apparent.

Materials

a) particulate sample collected on a Millipore filter using the volumetric sampling technique (Section 7.2). The filter should be stored in a petri dish until examination.

b) a clean 2 × 3 in. glass slide

c) stopcock grease

d) microscope

e) light source on adjustable mount

Procedure

a) Coat one side of the glass slide lightly and evenly with a thin film of grease.

b) Remove the sample filter from its storage dish. Place it, sample side up, on the greased surface of the glass slide. This insures that the filter paper will be flat for observation purposes. Do not use a cover slip. Any contaminant on it might interfere with observations. (If the filter is to be kept for future examination, place the top of a petri dish over the filter to protect it from further contamination.)

c) Adjust the incident light source to about 40° from the horizontal over the microscope stage (Fig. 7-3).

d) Examine the filter, first using low power and then a suitable higher power. Move the slide across the stage at intervals to survey the entire sample.

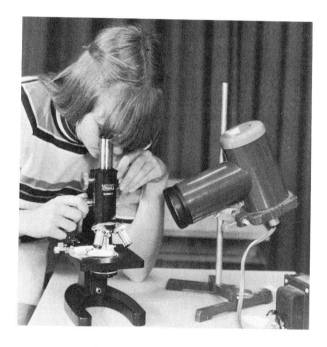

Fig. 7-3
Position of the incident light source relative to the microscope stage.

C. EXAMINATION USING TRANSMITTED LIGHT

Transmitted light is often best for viewing very small particles. When the Millipore filter is soaked with an immersion oil of suitable refractive index, it becomes quite transparent.

Fig. 7-4
Suitable immersion oil renders the sample filter transparent. (Courtesy of Millipore Corp.)

Materials

a)   particulate sample collected on Millipore filter
b)   immersion oil of refractive index 1.51
c)   petri dish
d)   forceps
e)   microscope with sub-stage illumination

Procedure

a)   Using forceps, carefully float the sample filter, sample side up, on a film of immersion oil in the cover of a petri dish (Fig. 7-4). Draw the bottom of the filter paper over the rim of the cover to scrape off any excess oil from the filter.

b)   Place the filter on a glass slide and examine it with the microscope. Start with low power and increase the magnification after you have surveyed most of the filter surface.

Discussion

How many different types of particles were you able to identify? How do different samples from different test sites compare with respect to the type of particles and the relative distribution of particle size? Can you suggest the most probable sources of these particles from your inspection of the samples?

7.4   DETERMINATION OF ORGANIC AND INORGANIC COMPONENTS OF AIRBORNE PARTICULATE MATTER

Air may contain organic solids such as soot, pollen grains, and rubber (from automobile tires). It may also contain many inorganic solids such as salt, metal fragments, and asbestos fibers. The sample filter from Section 7.2 probably holds both organic and inorganic solids. If you investigate particulate samples collected at several different sites, you should be able to detect a relationship between the relative proportions of organic and inorganic solids and the probable nature of their sources. For example, the organic fraction will likely be high near a coal-burning site, particularly if the combustion process is incomplete. The inorganic fraction may be high near a construction site where metals, cement, and insulating materials are being handled.

Materials

a) the sample filter from Section 7.2. This sample should be stored in a petri dish until tested.

b) standard porcelain crucible with cover

c) Bunsen burner

d) ring clamp; ring stand; clay triangle

e) ethanol

f) muffle furnace, capable of heating to 750°C

Procedure*

The weight of the inorganic fraction of the particulate sample is determined by burning off the organic portion.

a) Clean a small porcelain crucible and cover.

b) Place the clay triangle on the ring clamp attached to a ring stand.

c) Place the crucible with cover in the clay triangle and then heat it strongly with the Bunsen burner for 10 minutes. This will vaporize any surface contamination which might alter experimental values.

d) When it has cooled to room temperature, weigh the covered crucible to the nearest 0.1 mg. Record this value.

e) Place the sample filter in the crucible. Wet the filter with ethanol and then ignite it.

f) Cover the crucible and place it in the muffle furnace set at 750°C for 20 minutes.

g) Remove the crucible from the furnace and let it cool to room temperature.

h) Reweigh the covered crucible and contents, recording the final value to the nearest 0.1 mg.

*Provided through the courtesy of Millipore Corporation, Bedford, Mass. Copyright ® Millipore Corporation (1963).

Calculations

a)   Final weight of covered crucible + contents
        − Initial weight of covered crucible
            Weight of total inorganic content

b)   The inorganic contents of the crucible include the inorganic component of the Millipore filter paper. All of the organic portion will have been ignited. The weight of this filter ash is recorded on the Millipore package. This must be subtracted from the weight calculated in a).

c)   Calculate the percentage of inorganic solids in the sample as follows:

$$\frac{\text{Weight in mg of inorganic component of sample}}{\text{Weight in mg of total sample (Section 7.2)}} \times 100$$

Discussion

Do the relative proportions of organic and inorganic fractions vary when samples from different testing sites are analyzed? If so, explain the variation.

7.5   CHEMICAL IDENTIFICATION OF INORGANIC COMPONENTS OF PARTICULATE MATTER

Chemical spot testing provides a fast and extremely sensitive detection of elements such as lead which may be present in your particulate sample. Airborne particulate matter collected by filtration is deposited uniformly over the filter surface. Positive chemical spot tests produce a color reaction on the paper surface. The quantity of element which must be present in the sample before positive identification can be made varies; the average minimum value is approximately 10^{-8} grams.

     It is extremely important that only distilled or demineralized water be used in the preparation of reagents. Tap water can contain enough dissolved metal to yield false results.

A. SOLUBILIZING THE SAMPLE*

The material to be identified must be partially or totally dissolved. This involves treatment of the sample with an acid or mixture of acids.

*Provided through the courtesy of Millipore Corporation, Bedford, Mass. Copyright® Millipore Corporation (1963).

Materials

a) Millipore 47 mm Microfiber Glass prefilter

b) Millipore plastic petri dish

c) lacquered or teflon-coated forceps (Ordinary metal forceps will be attacked by the acid and may interfere with the test.)

d) 1.5 ml of 6*M* hydrochloric acid

e) 1.5 ml of 6*M* nitric acid

f) Millipore filter paper containing collected sample of particulate matter

Procedure

a) Place a 47 mm Microfiber Glass prefilter in a plastic petri dish. Add 1.5 ml *each* of 6*M* hydrochloric and nitric acids to the petri dish to saturate the prefilter pad. (Note: For the nickel test, use only hydrochloric acid.)

b) Place the test filter, *sample side up*, on the acid-soaked prefilter. Cover the petri dish.

c) Allow to stand for 10 minutes. Dissolving can be aided by placing the petri dish in a 60°C oven during this period.

d) Remove the filter. Place it on a 2 × 3 in. glass slide and allow to dry.

e) If the sample is to be tested for a number of different elements, the acid-treated filter should be cut into the appropriate number of wedge-shaped segments (Fig. 7-5).

Fig. 7-5
Each segment of an acid treated filter can be tested for a different element. (Courtesy of Millipore Corp.)

Each segment can then be tested for a different element. (The cutting blade used may leave enough contaminant at the edge of a segment to yield a positive test, but this will not extend into the central portion of the segment.)

B. IDENTIFICATION OF LEAD

Lead forms a red crystalline precipitate when tested with tetrahydroxyquinone (THQ). By obtaining a particle count of the crystals formed on the filter, the lead content of sample collected at different sites can be determined and compared. This test has been used in many areas to determine the lead content of exhaust fumes from automobile engines.

Materials

a)   tetrahydroxyquinone reagent (THQ). For best results, this should be prepared fresh daily. Since it is used to soak the prefilter pad, the amount required depends upon the number of samples to be tested. Add excess solid tetrahydroxyquinone to an appropriate volume of isopropanol. Allow it to dissolve until the solvent is saturated. Filter the solution and dilute the filtrate with an equal volume of distilled water.

b)   Millipore Microfiber Glass prefilter pad

c)   Millipore plastic petri dish

d)   teflon-coated forceps

e)   solubilized sample filter paper

Procedure

a)   Soak a Microfiber Glass prefilter pad with the THQ reagent.

b)   Place the solubilized sample filter, *sample side up*, on the pad. If lead is present, the filter will turn red.

C. IDENTIFICATION OF IRON

The reagent potassium ferrocyanide reacts with iron(III) salts to produce Prussian Blue, $Fe_4[Fe(CN)_6]_3$. This is a very sensitive test for iron. The results are sometimes masked by the presence of large quantities of copper or molybdenum, but this problem is rare.

Materials

a)   a saturated aqueous solution of potassium ferrocyanide. Only a drop is required for each test.

b)   solubilized sample filter paper

Procedure

a)   While the filter is still slightly damp from the solubilizing process, add a drop of the potassium ferrocyanide solution. A vivid blue color indicates the presence of iron.

b)   An alternate test uses only hydrochloric acid in the solubilizing process. A drop of 1% solution of potassium thiocyanate is then applied to the filter. If iron(III) ion is present, a red color will appear.

D. IDENTIFICATION OF NICKEL

Dimethylglyoxime produces a bright red insoluble salt with nickel ion. This salt, however, will not precipitate in the presence of nitrates and strong oxidizing reagents. Therefore the sample to be tested should be solubilized using only hydrochloric acid. Other metal salts which can interfere with the desired reaction are treated before the addition of dimethylglyoxime.

Materials

a)   sodium tartrate, saturated aqueous solution, 1 drop per test

b)   sodium carbonate, saturated aqueous solution, 1 drop per test

c)   dimethylglyoxime, 1% solution in ethyl alcohol, 1 drop per test

d)   ammonium hydroxide, 6M, 1 drop per test

e)   sample filter solubilized with hydrochloric acid only

Procedure

a)   Add one drop each of sodium tartrate and sodium carbonate solution to the dry, solubilized sample filter in order to react any iron present.

b)   Add one drop each of dimethylglyoxime and ammonium hydroxide. The development of a red color indicates nickel is present. This red nickel precipitate is not stable and will fade within 10 to 15 minutes.

E. IDENTIFICATION OF COPPER

This is a highly sensitive test for copper. Since a trace of copper is sometimes present in distilled water, a blank should be run periodically if the tests are persistently positive.

Materials

a)   rubeanic acid (dithiooxamide), 1% solution in ethanol, 1 drop per test

b)   malonic acid, 20% aqueous solution, 1 drop per test

c)   ethylenediamine, 10% aqueous solution, 1 drop per test

d)   solubilized sample filter

Procedure

a)   If the presence of nickel or iron is known or suspected, add one drop each of malonic acid and ethylenediamine solution.

b)   Add one drop of rubeanic acid. If copper is present, a black-olive-green precipitate will form immediately.

7.6   A SURVEY OF GASEOUS POLLUTANTS

What kind of gaseous pollutants would you expect to find in the air which you are testing? Carbon monoxide? Nitrogen dioxide? Are these gases present in significant concentrations? You can "survey" the air around you with rapid and simple gas sampling detector tubes. Each tube contains an accurately measured amount of fine grain silica gel. The gel is impregnated with a chemical which can detect the specific gas being analyzed. When an air sample containing the gas is drawn through the detector tube, a color change is observed. The length of the color stain is proportional to both the concentration of the gas and the length of time it is exposed to the indicator gel. For some tests the degree of color change is the critical factor. The method outlined is designed for the gas sampling tubes produced by Eduquip. These tubes are inexpensive and readily available.

Materials

a)   Eduquip Gas Sampling Kit. Includes syringe, rubber adapter, color code chart, and detection tubes for carbon monoxide, sulfur dioxide, nitrogen dioxide. (Tubes are also available for carbon dioxide, oxygen, and hydrogen sulfide.)

b)   Millipore hand vacuum assembly; rubber tubing for attachment. At low gas concentrations, large volumes must be tested at a flow rate of 100 ml per minute.

c)   timing device, for collection of large air samples

d)   cigarettes

e)   Millipore Sterifil filter holder

f)   rubber tubing, 2 in.

Procedure

a)   Using a suitable tool, *carefully* break the tip off each end of the selected gas detector tube.

b)   Carefully insert one end of the tube into the rubber adapter. Insert the syringe tip into the other end of the rubber adapter.

c)   Place the open detector tube end into the air sample to be studied.

d)   Pull the syringe plunger steadily back to the 25 ml mark. Hold it until the pressure inside the syringe stabilizes.

e)   If no color change is detected after a 2 minute period, repeat step d) at 2 minute intervals until the indicator changes color.

f)   Unless you are testing near an obvious source of the gas, the concentration should be low. A large air sample will be required for detection purposes. In this case the Millipore hand vacuum should be attached to the indicator tube in place of the syringe. Draw the air sample through at a rate of 100 ml per minute until the indicator changes color. *Record the total testing time.*

g)   Record the total volume of air passed through the gas detector tube. Match the indicator tube to the color chart. Record the corresponding concentration listed on the chart.

h)   The concentrations listed on the chart are based on air sample volumes of 100 ml. Determine the mathematical factor by which the experimental air sample volume would be increased or decreased to convert to 100 ml. Multiply the concentration from the chart by this factor. (For example, experimental volume = 50 ml; multiply chart concentration by 2. Experimental volume = 400 ml; multiply chart concentration by ¼ or 0.25.)

i) Record the calculated concentration of the gas tested.

Sulfur Dioxide

a) Determine the concentration of sulfur dioxide close to major sources of fuel combustion. Try to select sites using different types of fuel such as coal, oil, or natural gas. Test while these sources are in operation.

b) Determine the sulfur dioxide level in the atmosphere outside your home or school. This test may require a large air sample.

c) Draw suitable conclusions from your results.

Carbon Monoxide

a) Determine the concentration of carbon monoxide in an area of high traffic density or in a site where automobile exhaust cannot disperse readily (underground parking lot, traffic tunnel, service station). Try to test such sites during peak periods of activity.

b) Sample the carbon monoxide emission of an automobile in a well-ventilated area. If possible, compare the exhaust of 2 different automobiles—one with exhaust emission controls, the other without. Does altering the idle speed affect your results?

c) Test for carbon monoxide emission near sources of fuel combustion. What factors will affect your results?

d) Determine the carbon monoxide emission from a burning cigarette. The assembly used to collect the smoke is shown in Figure 7-6. Record the total volume of air passed through the Sterifil system. To test the carbon monoxide concentration, remove one of the rubber caps from the top of the filter holder. Then insert the detector tube attached to the syringe. Since a very small sample of the collected emission may yield a color change in the tube, pull the syringe plunger back gradually, checking at small intervals. Using the color chart, calculate the concentration of carbon monoxide in 100 ml of sample.

Fig. 7-6
The Sterifil system provides an excellent smoke collection assembly.
(Courtesy of Millipore Corp.)

e)   "Smoke" one or two cigarettes in a closed car using the Millipore hand vacuum assembly. Determine the concentration of carbon monoxide in the car. Convert your results to ppm. Would this concentration of carbon monoxide have any affect on the driver of the car? (Consult Section 4.5.)

Nitrogen Dioxide

a)   Determine the concentration of nitrogen dioxide close to major fuel combustion operations. As with the sulfur dioxide investigation, try to select sites burning different types of fuels. Which type of fuel would you expect to emit the greatest proportion of nitrogen oxides during combustion? Why?

b)   In a well-ventilated area, investigate the nitrogen dioxide emission from automobile exhaust. If possible, compare the exhaust concentrations of diesel and gasoline-powered vehicles. You might set up operations at the local bus stop or terminal. Which type of engine should yield the greatest concentration of nitrogen dioxide? Why? Do your results confirm your prediction? You might also test the latest car models, equipped with emission control devices.

Additional Studies

Reread Section 1.4. Predict the relative amounts of oxygen and carbon dioxide at a number of sites—your classroom, the schoolyard, near dense vegetation on a sunny day, near a source of combustion. Use Eduquip detection tubes to check your predictions.

Reread Section 4.4. Select sites where the hydrogen sulfide concentration might be high. Determine the concentrations using Eduquip detection tubes.

7.7   TOTAL ACIDS IN AIR

Elimination of air pollution is a costly enterprise. However, the average citizen pays a great deal more to repair the extensive damage caused by air pollutants. A New York City resident may spend as much as $200 yearly to clean, wash, repaint, repair, and even replace material possessions exposed to the air. And who can place a price tag on human health?

Much of this damage is produced by the corrosive action of acids. Acids form when certain contaminants, acidic in nature, encounter moisture. For example, both sulfur dioxide and nitrogen dioxide produce an irritating acid when inhaled into the moist respiratory tract.

How acidic is the atmosphere in your community? In this exercise, you will collect both particulate and gaseous pollutants and then determine the concentration of contaminants which are acidic in nature. For measurable results, you must select sites where pollution emission is significant. Refer to Section 7.6 for recommended test areas.

A. SAMPLE COLLECTION

Materials

a) long-stem glass funnel
b) large side-arm test tube with 1-hole rubber stopper
c) distilled water
d) rubber tubing
e) Millipore aerosol adapter; set of Millipore limiting orifices
f) vacuum source. For sample collection in the field, a hand vacuum cleaner that plugs into the lighter of a car is convenient. *The vacuum source selected must be capable of drawing air through the limiting orifice at the specified rate.*
g) thermometer
h) barometer
i) stoppered, 250 ml Erlenmeyer flask for each sample
j) wash bottle
k) waterproof marker

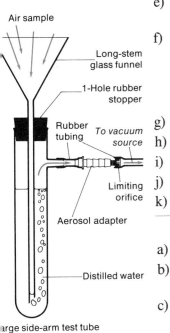

Air sample

Long-stem glass funnel

1-Hole rubber stopper

Rubber tubing   To vacuum source

Limiting orifice

Aerosol adapter

Distilled water

Large side-arm test tube

Fig. 7-7
Apparatus for gas sample collection.

Procedure

a) Assemble the collection apparatus shown in Figure 7-7.
b) Measure and record the temperature and pressure of the air being sampled.
c) The sampling time and rate are dependent on the general air quality at the testing site. Refer to the guide in Section 7.2. During the collection, record both the rate and the total sampling time.
d) When collection is complete, rinse any particulate matter adhering to the inside of the funnel into the Erlenmeyer

flask. Use *small* portions of distilled water from the wash bottle.

e) Carefully pour the solution from the test tube into a clean, labeled 250 ml Erlenmeyer flask. Rinse the sides of the test tube with 2–3 *small* portions of distilled water from the wash bottle. Carefully add these rinsings to the Erlenmeyer flask. Do not dilute the total sample with excessive rinsing. Stopper the sample flask until the contents can be tested.

B. SAMPLE TESTING

Materials

a) ring stand
b) buret clamp
c) buret with meniscus reader
d) $0.01M$ sodium hydroxide solution
e) phenolphthalein indicator solution
f) Erlenmeyer flask containing dissolved sample

Procedure

The concentration of acid can be determined by neutralization with sodium hydroxide. When sufficient sodium hydroxide has been added to neutralize the solution, the phenolphthalein indicator turns from colorless to pink.

a) Rinse the inside, including the tip, of a clean buret with a small portion of $0.01M$ sodium hydroxide solution. Drain the buret.

b) Fill the buret with $0.01M$ sodium hydroxide solution. Mount it on a buret clamp attached to the ring stand. Drain off a few drops of the sodium hydroxide to insure that the delivery tip is completely filled with solution. Record the initial volume of sodium hydroxide solution in the buret.

c) Add 2–3 drops of phenolphthalein indicator solution to the contents of the Erlenmeyer flask.

d) Position the sample flask for titration (Fig. 7-8). A white surface placed under the flask aids in seeing the color change.

e) Gently swirl the contents of the flask while adding sodium hydroxide *dropwise*. This permits better mixing of

Fig. 7-8
Gently swirl the flask co tents during titration permit better mixing of th reactants.

the reactants. After each drop a faint pink color will likely appear in the region where the drop landed. This color results from the temporarily high concentration of sodium hydroxide. Swirl until this color disappears before adding another drop. As the concentration of sodium hydroxide in the solution increases, the pink color will fade more slowly. When all of the acid has been neutralized, the addition of a single drop of sodium hydroxide solution will produce a permanent, faint pink color throughout the solution. Discontinue the addition of sodium hydroxide. Record the final volume in the buret.

Neutralization

An acidic solution contains an ion species represented by H_3O^+. A basic solution such as sodium hydroxide contains the ion species OH^-. When these ions are mixed, a *neutralization* reaction occurs:

$$H_3O^+ \quad + \quad OH^- \quad \rightarrow \quad 2\,H_2O$$
acidic species   basic species   water

The actual size and mass of an ion defies the imagination. In order to work with measurable quantities of ions, chemists have devised a unit called the *mole*. A mole, like a dozen, represents a certain *number*, 6.02×10^{23}. For example, 1 mole of water contains 6.02×10^{23} water molecules. One mole of an acid solution can produce 6.02×10^{23} H_3O^+ ions. Similarly 1 mole of a basic solution can produce 6.02×10^{23} OH^- ions. Since 1 OH^- ion is required to neutralize 1 H_3O^+ ion, 1 mole of OH^- ions is required to neutralize 1 mole of H_3O^+ ions. Hence, the number of moles of acid in the sample solution can be determined by calculating the number of moles of sodium hydroxide base which were added to neutralize the acid.

Calculations

a)   Volume of $0.01\,M$ sodium hydroxide added to sample $=$
        initial volume reading (ml) $-$ final volume reading (ml)

b)   moles of OH^- added $=$ moles of sodium hydroxide added
    $=$ volume of sodium hydroxide added (ml)

$$\times \frac{0.01 \text{ moles of sodium hydroxide}}{1{,}000 \text{ ml sodium hydroxide}}$$

$=$ moles of acid in sample solution

c) The number of moles of acid determined in calculation b) was originally contained in the volume of air which was drawn through the collection apparatus. Calculate the volume of air according to the following:

$$\text{rate of collection} \left(\frac{\text{liters}}{\text{minute}}\right) \times \text{total collection time (minutes)}$$

d) Determine the concentration of acid in the air tested:

$$\frac{\text{concentration}}{\text{of acid}} = \frac{\text{moles acid collected}}{\text{volume of air sampled (liters)}}$$

$$= \frac{\text{moles of acid}}{1 \text{ liter of air}}$$

e) The acidity of air is generally expressed in ppm by weight. However, in order to determine the *weight* of acid collected, the actual composition of the acid solution would have to be determined by complex analysis. To greatly simplify this procedure and still obtain a measure of acid content in ppm, *assume* that the basic component of the acid solution is sulfuric acid, H_2SO_4. Since the sulfur dioxide (SO_2) which produces this acid is a major pollutant in most urban areas, this assumption is likely to be valid. One mole of acid is produced from 49.0 gm of H_2SO_4. Calculate the number of grams of H_2SO_4 required to produce the number of moles of acid found in 1 liter of the air sample (step d).

f) Knowing the volume of air sampled, you can calculate its weight from the density of the air at the experimental temperature and pressure. Use the following formula to calculate the density of the air sample:

$$D_2 = D_1 \times \frac{P_2}{P_1} \times \frac{T_1}{T_2}$$

where: D_2 = density of air at the recorded experimental pressure and temperature in gm/l
D_1 = 1.29 gm/l (density of air at standard temperature and pressure, STP)
P_2 = experimental pressure (mm)
P_1 = 760 mm (standard pressure)
T_1 = 273°K (standard temperature)
T_2 = experimental temperature (°C) + 273K°
= experimental temperature (°K)

g) weight (gm) of the air sample =
    density of air sample (D_2, gm/l) × volume of air sample (l)

h) The relative weight of acid (H_2SO_4) in the air sample is given by:

$$\frac{\text{weight (gm) of acid } (H_2SO_4)}{\text{weight (gm) of the air sample}} = \frac{\text{weight (gm) of acid } (H_2SO_4)}{1 \text{ gm of the air sample}}$$

i) Determine the concentration of acid in ppm by weight by calculating the weight (gm) of acid (H_2SO_4) in 1×10^6 gm of the air sample.

Discussion

Does the acidity of the air sampled at the various test sites vary significantly? Explain your results.

What are the major sources of acidic pollutants in your area? In view of this, were you justified in assuming that the primary acid component was sulfuric acid, resulting from sulfur dioxide emissions? If not, which other acids were you more likely to collect?

7.8 SULFUR DIOXIDE IN AIR

Large sources of sulfur oxide emissions are concentrated in specific regions of North America. For example, the industrial northeastern states rely heavily on coal and residual fuel oil for heat and power. These states contribute almost 50% of the total yearly emission of sulfur oxides in the United States. Arizona and Texas are also high-ranking offenders. Smelters and refineries account for more than 80% of the total sulfur oxides emission in each state. A similar situation exists in the Sudbury mining region of Ontario.

Is sulfur dioxide a major pollutant in your area? Testing sites must be selected near large sources of sulfur oxide emissions before sufficient concentrations can be obtained for this analysis. In the following procedure, airborne sulfur dioxide (SO_2) is collected and dissolved in distilled water. The concentration of the resulting acid solution is determined by titration. From this analysis, the actual concentration of airborne sulfur dioxide can be calculated.

A. SAMPLE COLLECTION

Refer to Section 7.7 for the required sampling materials and procedure. For significant results, try to collect as large an air sample as time permits.

B. SAMPLE TESTING

Materials

a) ring stand
b) buret clamp
c) buret with meniscus reader
d) 0.01M standardized potassium permanganate solution. For accurate results, this solution should be standardized against a sodium oxalate solution of known concentration just prior to use. (The concentration of the potassium permanganate solution can change significantly during a 24-hour period.)
e) Erlenmeyer flask containing dissolved gas sample
f) wash bottle with distilled water
g) dilute sodium hydrogen sulfite solution—for rinsing the buret after titration.

Procedure

Sulfur dioxide dissolves in water to produce sulfurous acid according to the following equation:

$$\text{Sulfur Dioxide} + \text{Water} \rightarrow \text{Sulfurous Acid}$$
$$SO_2 + H_2O \rightarrow H_2SO_3$$

When this acid solution is titrated with potassium permanganate solution, the following reaction occurs:

$$\begin{array}{l}\text{Sulfurous} + \text{Potassium} \rightarrow \\ \quad \text{Acid} \quad \text{Permanganate}\end{array}$$

$$\begin{array}{l}\text{Potassium} + \text{Manganese} + \text{Sulfuric} + \text{Water} \\ \text{Sulfate} \qquad \text{Sulfate} \qquad \text{Acid}\end{array}$$

$$5\,H_2SO_3 + \quad 2\,KMnO_4 \quad \rightarrow$$

$$K_2SO_4 \quad + \quad 2\,MnSO_4 \quad + 2\,H_2SO_4 + 3\,H_2O$$

The dark purple permanganate solution is reduced to colorless reaction products. The reaction is complete when all of the sulfurous acid has been reacted. Further addition of potassium permanganate solution results in a permanent purple color throughout the sample solution.

a) Rinse the inside, including the tip, of a clean buret with a small portion of $0.01M$ potassium permanganate solution. Drain the buret.

b) Fill the buret with $0.01M$ potassium permanganate solution. Mount it on the buret clamp attached to the ring stand. Collect a few drops of the solution in a beaker to ensure that the delivery tip is completely filled.

c) Record the initial volume of potassium permanganate solution in the buret. Position the sample flask for titration. Place a white surface under the reaction flask to aid in the detection of color change.

d) Gently swirl the contents of the flask while adding potassium permanganate solution *dropwise*. This agitation permits better mixing of the reactants. After each drop a faint purple color will likely appear in the region where the drop landed. This color is produced by the temporarily high concentration of potassium permanganate. Swirl until this color disappears before adding another drop. As the concentration of potassium permanganate in the solution increases, the purple color will fade more slowly. Finally, when the acid has completely reacted, the addition of a single drop will produce a permanent purple color throughout the solution. Discontinue the addition of potassium permanganate and record the final volume in the buret.

e) The potassium permanganate solution may leave a brown stain on the sides of the buret. Add a few drops of sodium hydrogen sulfite solution ($NaHSO_3$) to the rinse water to remove this deposit. Then rinse the buret well with water.

Calculations

The reaction equation shows that 2 molecules of potassium permanganate ($KMnO_4$) react with 5 molecules of sulfurous acid (H_2SO_3). Or, in the language of the chemist, 2 *moles* of $KMnO_4$ react with 5 *moles* of H_2SO_3. (Refer to *Neutralization*—Section 7.7.) By calculating the number of moles of $KMnO_4$ which were added to react all of the acid, the number of moles of sulfur dioxide (SO_2) which produced the acid can be determined.

a) Volume of potassium permanganate added to acid = initial volume reading (ml) − final volume reading (ml).

b)   moles of potassium permanganate added to the acid =

volume of potassium permanganate (ml) ×

$$\frac{0.01 \text{ moles of potassium permanganate}}{1,000 \text{ ml of potassium permanganate}}$$

c)   2 moles of potassium permanganate react with 5 moles of sulfurous acid. Therefore, 1 mole of potassium permanganate reacts with 5/2 moles of sulfurous acid. Calculate the number of moles of sulfurous acid which reacted with the number of moles of potassium permanganate calculated in b).

d)   1 mole of sulfur dioxide (SO_2) produces 1 mole of sulfurous acid when dissolved in water. Therefore:

moles of SO_2 dissolved = moles of SO_2 in air sample
                                        = moles of sulfurous acid calculated in c)

e)   The total volume of air which contained the SO_2 is given by:

$$\text{rate of collection} \left(\frac{\text{liters}}{\text{minute}}\right) \times \text{total collection time (minutes)}$$

f)   Determine the concentration of SO_2 in the air sample:

$$\frac{\text{concentration}}{\text{of } SO_2} = \frac{\text{moles of } SO_2 \text{ collected}}{\text{volume (l) of air sampled}} = \frac{\text{moles of } SO_2}{1 \text{ liter of air}}$$

g)   The sulfur dioxide concentration is usually calculated in parts per million (ppm) by volume. The volume occupied by a mole of gas molecules depends upon both the temperature and the pressure of the gas. Equal volumes of gases measured at the same temperature and pressure contain equal numbers of molecules. Furthermore, one mole of any kind of gas molecules, measured at standard temperature and pressure (STP), occupies 22.4 liters. If 1 mole of sulfur dioxide gas (SO_2) occupies 22.4 l at STP, calculate the volume occupied at STP by the number of moles of SO_2 in the air sample.

h)   Use the following formula to convert this calculated volume of SO_2 at STP to the actual volume of SO_2 collected under the experimental conditions of temperature and pressure:

$$V_2 = V_1 \times \frac{P_1}{P_2} \times \frac{T_2}{T_1}$$

where: V_2 = volume (l) of SO_2 at experimental temperature and pressure

V_1 = volume (l) of SO_2 at STP [calculated in g)]

P_1 = 760 mm (standard pressure)

P_2 = experimental pressure (mm)

T_1 = 273°K (standard temperature)

T_2 = experimental temperature (°C) + 273K°

= experimental temperature (°K)

i)   Relative volume of SO_2 in the air sample =

$$\frac{\text{volume (l) of } SO_2 \text{ in the air sample}}{\text{volume (l) of air sample}} = \frac{\text{volume (l) } SO_2}{1 \text{ liter of air sample}}$$

j)   Determine the concentration of SO_2 in ppm by volume by calculating the volume (l) of SO_2 in 1×10^6 liters of the air sample.

Discussion

Does the concentration of sulfur dioxide in the air sampled at the various test sites vary significantly? Explain your results in view of the suspected sources.

Explain any changes which you might expect to measure in the level of sulfur dioxide in the air in your community during a 24-hour period; an average week; and at different seasons of the year.

Compare your experimental results with the standards governing sulfur dioxide levels in your community. If these data are not available, refer to Section 4.4.

7.9   EFFECTS OF AIR POLLUTANTS

Unit 4 outlined many of the effects of airborne pollutants on living organisms and non-living materials. You can design your own investigations of some of these effects. Following the suggestions outlined here, select the gases to be tested, the materials to be exposed, and also the experimental conditions. The time period required will depend upon the concentration of the gaseous pollutant and on the type of material exposed.

The gases used in these studies are highly toxic. DO NOT INHALE ANY OF THEM. USE A FUME HOOD.

Test Chambers

The investigations should be performed in airtight test chambers. Test chambers of any dimensions can be constructed using panes of glass purchased from a local hardware store. Tape all the sharp edges and seal the panes together with epoxy glue. The design must obviously allow for the introduction of test materials and gases. Wide-mouth glass jars, with reliable screw tops, are very convenient for small-scale studies. Test samples can be suspended from the inner lid using fine string and tape (Fig. 7-9). Or, glass aquaria and large beakers can be inverted and secured on a cardboard or wooden base. Bell jars are also easy-to-use test chambers.

Gaseous Pollutants

Unless an airtight gas inlet has been designed, test chambers should *always* be placed under a fume hood while gases are being introduced. The concentration of the gas in the test chamber can be determined using Eduquip gas sampling detector tubes.

Gases such as sulfur dioxide (SO_2), nitrogen dioxide (NO_2), and ozone (O_3) are available in cylinders from many sources. They can also be generated using standard laboratory methods. Ask your instructor for a reference text.

Test Materials

A wide range of materials should be exposed to the gaseous pollutants, either in a test chamber or in the existing atmosphere in your community. The extent to which any material will be affected depends upon the nature of the material; the duration of exposure to gaseous pollutants; and the concentration of the pollutant.

Fig. 7-9
Samples to be tested are suspended from the inner lid of a wide-mouth glass jar.

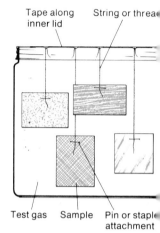

Tape along inner lid   String or thread

Test gas   Sample   Pin or staple attachment

Procedure

The following investigations are suggested. Remember to use *controls.*

Non-living Materials

a) Test a number of metals and metal alloys to determine the degree of their resistance to pollutants. Examples are copper, brass, aluminum, chromium, silver, tin, zinc, and stainless steel.

b) Find out whether stretched or unstretched rubber bands are more sensitive to ozone. Examine the test specimens regularly to determine the degree of cracking and loss of elasticity.

c) Test different fabrics—cotton, nylon, silk, wool, and synthetics—in a variety of colors. Dyes vary in resistance to airborne oxidants. Microscopic examination should reveal broken threads or other signs of deterioration.

d) Test the resistance of different types of paints, stains, and varnishes. Apply each product to small sample areas which are then exposed to pollutants.

Living Organisms

Air pollution severely damages many types of vegetation. Gaseous contaminants generally enter a plant through the stomata (pores) on the leaves. Plant species differ greatly in their sensitivity to various pollutants. For each gas tested, try to determine the maximum tolerance level for each group of plants. Record the nature and the rate of damage to the vegetation. Remember to provide the normal plant requirements and to set up controls.

a) To investigate sulfur dioxide damage, select plants from each of the following three groups. At the end of the experiment, categorize the groups as highly sensitive, moderately sensitive, or resistant.

Group 1: boxelder, chrysanthemum, corn, Irish potato, lilac (common or Persian), onion, rose, Virginia creeper.

Group 2: alfalfa, barley, bean, broccoli, brussel sprouts, carrot, chicory, clover, cosmos, cotton, endive, lettuce, lichens, oats, pumpkin, radish, rhubarb, salvia, salsify, spinach, squash, sweet potato, Swiss chard, table beets, tobacco, turnip, wheat.

Group 3: apple, apricot, aster, begonia, cabbage, cauliflower, cherry, gladiolus, grape, iris, marigold, parsley, parsnip, peach, pear, plum, rye, sugar beet, tomato, watermelon, zinnia.

b) To investigate ozone damage, select plants from each of the following three groups. Categorize the groups as in a). Repeat for nitrogen dioxide.

Group 1: barley, bean, egg plant, head lettuce, onion, parsley, radish, rhubarb, tomato, turnip, white pine.

Group 2: cabbage, cantaloup, cucumber, gladiolus, pumpkin.

Group 3: alfalfa, barley (young), celery, endive, oats, petunia, spinach, sugar beets, Swiss chard, table beets, tobacco.

c)   Investigate the effects of these gases on fungi such as mushrooms, bread mold, and yeast.

d)   Any investigations using animal life should be limited to small organisms such as fruit flies, mosquitoes, or ants.

Discussion

Which of the metals or metal alloys proved most resistant? Which fabric types were most readily damaged? What type of damage was most evident using different paint samples? How does sulfur dioxide damage to vegetation compare with that of ozone and nitrogen dioxide? Which plants are highly sensitive to all three gases? Why are some plants affected more than others? Does the age of the plant affect its sensitivity? Are plants affected to the same extent during the daily period of photosynthesis as during the hours of darkness? Compare the effects of these gases on chlorophyll-containing plants and fungi.

7.10   DEMONSTRATION OF A TEMPERATURE INVERSION

Under normal conditions, when the atmospheric temperature decreases rapidly with altitude, an air mass is unstable. Pollutants, released at ground level, are carried by vertical turbulence high into the cooler air above. However, when the air at or near ground level is cooler than the air above, a stable air mass is produced. While this temperature inversion prevails, vertical movement of the air mass is restricted. Contaminants emitted at ground level accumulate around their sources. The concentration of pollutants continues to build until the temperature inversion ends. The effects of a temperature inversion can be demonstrated with a model system.

Materials

a)   1 liter graduated cylinder

b)   2 or 3 liter beaker

c)   mixture of ice, salt, and water

d)   2 thermometers

e)   string or thread

f)   ring stand with thermometer clamp

g)   heat lamp with clamp

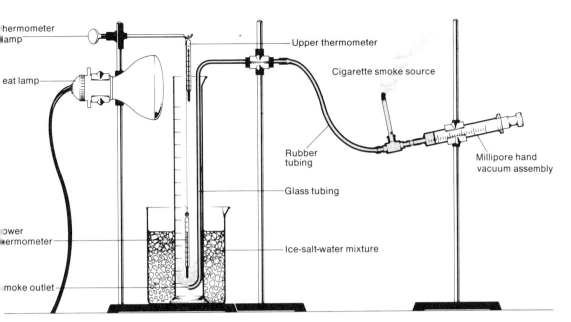

Labels in figure:
- thermometer clamp
- heat lamp
- power thermometer
- smoke outlet
- Upper thermometer
- Cigarette smoke source
- Rubber tubing
- Millipore hand vacuum assembly
- Glass tubing
- Ice-salt-water mixture

Fig. 7-10
Model demonstrating the effects of a temperature inversion.

h) glass tubing with elbow joint
i) rubber tubing
j) Millipore Hand Vacuum Assembly
k) cigarette

Procedure

a) Set up the apparatus as shown in Figure 7-10. Do not activate the smoke source or the heat lamp until the ice-salt-water mixture has cooled the air in the lower portion of the graduated cylinder.

b) Use the heat lamp to warm the air in the upper portion of the graduated cylinder.

c) When a significant temperature difference exists between the upper and lower regions of the cylinder, light the cigarette. Slowly introduce smoke into the bottom of the cylinder.

d) Observe the distribution of smoke within the cylinder.

e) After a dense layer of smoke has formed inside the cylinder, carefully remove the beaker containing the ice-salt-water mixture. Note the resulting temperature changes and the effect on smoke dispersal. (To speed this process, move the heat lamp closer to the base of the graduated cylinder.)

f) Demonstrate the dispersal of smoke under normal conditions. Use the heat lamp to warm the air in the bottom of the graduated cylinder to a higher temperature than the air in the upper portion.

Discussion

What is the minimum temperature difference necessary to maintain a stable layer of smoke in the graduated cylinder? Does the extent of the temperature difference between the upper and the lower regions of the cylinder greatly affect the upward dispersion of smoke? What natural situations were being simulated both prior to and following the removal of the ice-salt-water mixture?

7.11 ELECTROSTATIC PRECIPITATION OF DUST PARTICLES

Many industrial operations require the removal of solid particles from waste gases. This process not only reduces air pollution but often recovers valuable elements. For example, the flyash from the combustion of some British coals contains germanium. Useful potassium compounds are often recovered from the chimneys of cement factories.

One widely-used removal device is the electrostatic precipitator. A number of the particles emitted from any smoke stack carry an electric charge. A charge may be induced on the rest of the particles by passing them between two oppositely-charged electrodes. Positively-charged particles are drawn toward the negative electrode; negatively-charged particles are attracted to the positive electrode. The charge on each particle is neutralized when it reaches the attracting electrode. These neutral particles no longer repel one another, but simply collect. They form a solid mass which can no longer be supported by the flow of gases up the stack. Modern precipitators vary greatly in size. They can remove up to 99.9% of particles ranging in diameter from less than 0.1 μ to greater than 200 μ. The effectiveness of this device can be demonstrated with a model precipitator.

Materials
a)   500 ml graduated cylinder
b)   metal ring stand rod (or suitable substitute)
c)   large, 1-hole rubber stopper

d)   copper wire
e)   induction coil
f)   D.C. power source
g)   glass tubing with elbow joint
h)   Millipore Hand Vacuum Assembly
i)   rubber tubing
j)   cigarette

Procedure

a)   Assemble the apparatus shown in Figure 7-11. Note that the 2-way valve of the Millipore Hand Vacuum Assembly is attached in reverse to its usual order. It should pump smoke from the cigarette *into* the graduated cylinder.

b)   Light the cigarette and introduce smoke into the bottom of the graduated cylinder until it diffuses freely out of the top of the cylinder.

c)   Connect the induction coil and switch on the power. Observe the effect of electrical charge on the smoke particles in the cylinder. AVOID CONTACT WITH ANY OF THE ELECTRICAL TERMINALS WHILE THE VOLTAGE IS BEING APPLIED.

Fig. 7-11
Model of an electrostatic precipitator.

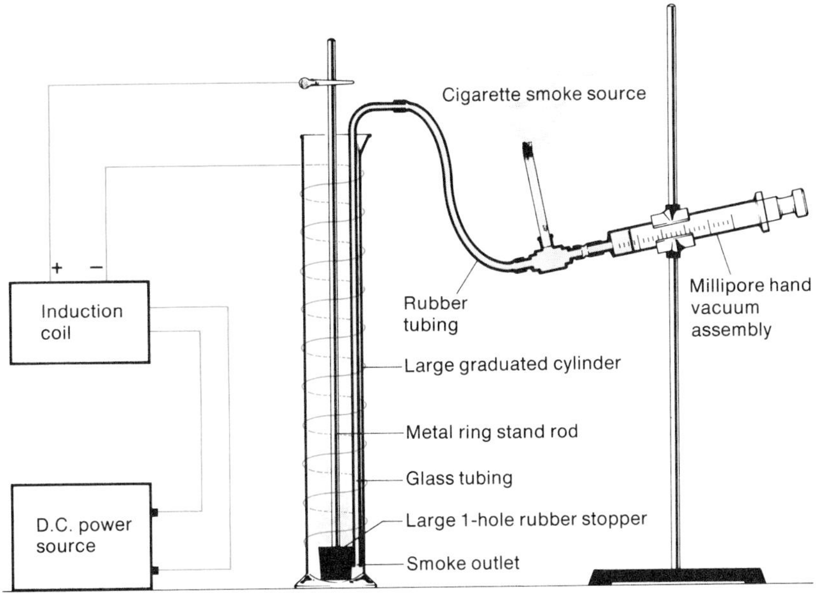

Cigarette smoke source

+   −

Induction coil

Millipore hand vacuum assembly

Rubber tubing

Large graduated cylinder

Metal ring stand rod

Glass tubing

D.C. power source

Large 1-hole rubber stopper

Smoke outlet

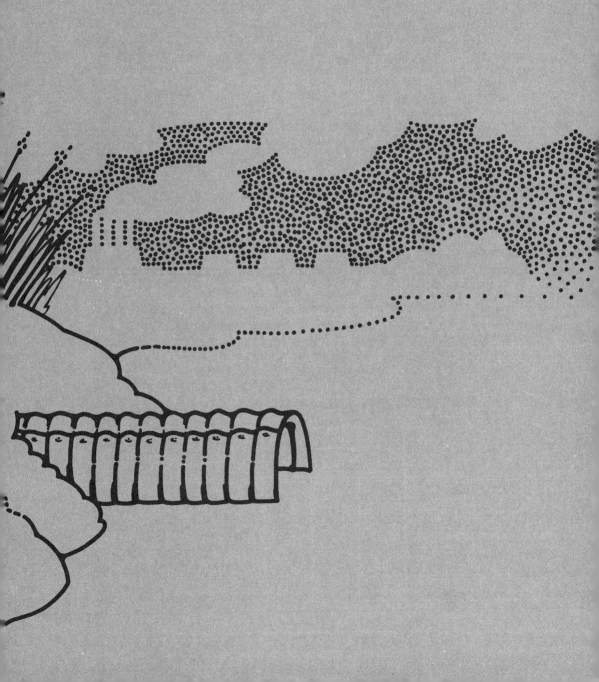

Case Studies

8

By now you should be quite aware of the interdependence of biotic (living) and abiotic (non-living) factors in an ecosystem. If any one of these factors is altered, most of the others also change. If, for example, the temperature of a body of water is changed, the D.O. changes and living organisms are affected. If first-order carnivores are harmed in some way, herbivores may increase and, consequently, producers may decrease. The chain of relationships is unending, complex but certain. You have undoubtedly discovered this in your field and laboratory work.

Pollution means more than just dirty air and dirty water with tin cans in it. By altering the environment, man threatens his very existence. It is quite amazing that a species which depends so heavily on air and water for its existence moves so slowly in putting a stop to pollution. Pollution is an inevitable result of civilization, but it can be controlled. If mankind fails to control pollution, pollution will eventually control mankind. Like any other species that destroys the environment on which it depends, the human species will decrease dramatically in numbers or become extinct if pollution continues to increase at its present rate.

Fig. 8-1

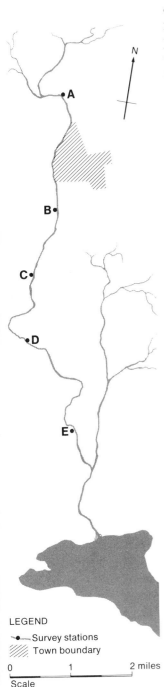

The knowledge that you have gained by working through this guide is of little value unless you can apply it. Unit 8 consists of case studies composed of actual data gathered by scientists from many parts of North America. By working through these, you can test your ability to apply your knowledge. Study them on your own and discuss your conclusions with others. But don't stop there! Continue to apply your knowledge to problems that you read and hear about. Read, think, and then act. Play the role of an informed and concerned citizen. Express your concern to the appropriate government officials and business executives.

8.1   EFFECT OF A LARGE TOWN ON A RIVER

The river in this study is 10 miles long and drains into a lake of moderate size. Situated on it is a town of about 15,000 people. A biological and chemical survey was performed to determine the effect of the town on the water quality of the river.

Five stations were set up along the river (Fig. 8-1). Station A was located about 0.5 mile above the town. Here the river is about 25 feet wide and 4 feet deep. Station B was located 0.5 mile downstream from the town. Here the river is also about 25 feet wide and 4 feet deep. Station C was located 1 mile further downstream. Station D was located 1.5 miles from C. The river at both C and D is about 20 feet wide and 3.5–4 feet deep. Station E was located two miles from D. The river is 18 feet wide and 3 feet deep there. The velocity of flow at all sampling stations was about 2.5 feet per second. The town has no sewage treatment plants. Its sewage is dumped untreated into the river. The results of the survey are tabulated in Tables 19 and 20.

TABLE 19   BOTTOM FAUNA PER SQUARE FOOT

	A	B	C	D	E
Mayfly nymphs	20	4	28	15	23
Stonefly nymphs	12	3	5	10	14
Caddisfly larvae	15	0	1	6	18
Asellus	7	5	6	8	7
Chironomus	2	26	24	21	6
Tubifex	1	37	36	24	8

TABLE 20 CHEMICAL ANALYSIS

	A	B	C	D	E
Suspended solids (ppm)	10	19	17	14	12
Phosphate (ppm)	0.37	0.75	0.61	0.43	0.41
B.O.D. (ppm)	1.8	3.2	3.1	2.6	2.1
Dissolved oxygen (ppm)	6.5	2.1	2.2	3.4	4.9
Nitrogen (ppm)	0.22	2.13	1.27	1.02	0.59
Coliforms per 100 ml	0	180	170	121	87

Questions

1 Account for the changes in the bottom fauna of the river.
2 Account for the high B.O.D. reading at station B.
3 What do the coliform counts tell you about the sanitary quality of the river?
4 What would you expect the relative populations of algae and zooplankton to be at the five stations?
5 Downstream from the town, the sensitive bottom fauna increase in numbers and the tolerant fauna decrease in numbers. Account for this change.
6 Use the chemical and biological data to predict the types of fish, if any, that might be found at each station.

8.2 EFFECT OF AN AGRICULTURAL AREA ON A RIVER

The river in this study passes through a relatively large agricultural area. A survey was made to determine the effects, if any, that agricultural practices have on the water quality of the river.

Five stations were set up along the river and one in the lake (Fig. 8-2). A chemical and biological analysis of the water was made at each station. Bottom fauna were studied; the results were tabulated in number per square foot. A relative study of the algae was also made at each station. The results are shown in Tables 21, 22, and 23.

The only significant algae present at stations B–E were *Cladophora*, *Spirogyra*, and a small quantity of *Ulothrix*. They were found only along the banks of the river. No significant quantities of phytoplankton were observed in the river. However, at station F there was a bloom of *Microcystis*. Also present, but in lesser quantities were *Navicula*, *Anabaena*, *Closterium*, and *Chlorella*.

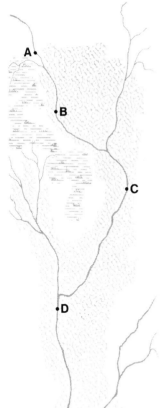

Fig. 8-2

TABLE 21 PHYSICAL CHARACTERISTICS

	A	B	C	D	E	F
Width (ft)	15	20	12	12	15	–
Depth (ft)	2	2.5	2	2	2.5	12
Velocity of flow (ft/sec)	3	1.5	2	2	2.5	–

TABLE 22 CHEMICAL ANALYSIS

	A	B	C	D	E	F
Suspended solids (ppm)	22	28	36	61	62	21
Phosphate (ppm)	0.04	0.09	0.16	0.75	1.1	2.9
B.O.D. (ppm)	1.9	2.2	2.0	1.8	1.9	2.8
Dissolved oxygen (ppm)	6.5	6.3	6.1	6.6	6.2	5.7
Nitrogen (ppm)	0.22	0.61	0.83	1.01	1.73	1.21

TABLE 23 BOTTOM FAUNA PER SQUARE FOOT

	A	B	C	D	E	F
Mayfly nymphs	16	12	13	7	1	0
Stonefly nymphs	9	8	6	2	0	0
Caddisfly larvae	13	11	7	1	4	1
Asellus	2	0	1	1	2	1
Chironomus	1	0	2	5	17	15
Tubifex	0	0	1	4	15	21

LEGEND

⸺ Swamp
•⸺ Sampling stations
    Agricultural area

0   1   2 miles
Scale

Questions

1 In general, how has the agricultural area affected the water quality of the river passing through it?

2 Account for the biological changes along the course of the river.

3 Would the numbers, kinds, and distribution of living organisms be any different if the velocity of flow averaged only 0.5 feet per second? Explain.

4 Account for the bloom of algae in the lake.

5 A study of the zooplankton was not made. What do you expect their distribution would be at these stations? Why?

6 Why is the B.O.D. higher in the lake than at any station in the river?

7 Why is the D.O. lower in the lake than at any station in the river?

8 Study carefully the changes in nitrogen and phosphate concentration. What might be responsible for these changes? Are these ions responsible for any of the biological changes in the waterway? Explain.

9 Account for the observed changes in suspended solids. Could the suspended solids be responsible for any of the biological changes? Explain.

8.3 EFFECT OF A CHEMICAL PLANT ON A RIVER

The river in this study passes by a chemical plant. A chemical and biological analysis of the water was made to determine the effect of the plant. Survey stations were set up as shown in Figure 8-3.

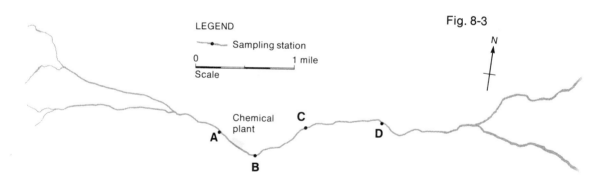

The velocity of flow at all stations was 1–1.5 feet per second. The river was 20–25 feet wide and 3–4 feet deep. At each station the bottom fauna were sampled at each bank and in the middle. The value in the table is an average of these three values.

The fish were sampled with a seine net. Four sweeps were made and the number of fish totalled. Two samples were tested at each station for coliforms; the values were averaged. The relative abundance of phytoplankton at each station was determined. The results of these studies are shown in Tables 24, 25, 26, 27, and 28.

TABLE 24 BOTTOM FAUNA PER SQUARE FOOT

	A	B	C	D
Mayfly nymphs	14	0	2	11
Caddisfly larvae	7	0	1	8
Stonefly nymphs	5	0	0	1
Chironomus	1	69	54	14
Tubifex	0	80	65	20
Dragonfly nymphs	4	0	0	1

TABLE 25 CHEMICAL ANALYSIS

	A	B	C	D
B.O.D. (ppm)	1.4	3.2	2.7	2.1
Phosphate (ppm)	0.8	1.2	1.1	1.0
D.O. (ppm)	6.4	2.1	2.9	5.8
Nitrogen, organic (ppm)	0.9	19.3	12.4	6.2
Nitrogen, ammonia (ppm)	0.37	15.01	12.23	7.19

TABLE 26

	A	B	C	D
Coliforms per 100 ml	0	3	2	0

TABLE 27 FISH CATCH

	A	B	C	D
Shiners	5	0	0	0
Crappies	10	0	0	2
Bluntnose minnow	2	0	1	2
Brown bullhead	7	0	1	0
Perch	2	0	0	3
Rockbass	2	0	0	0
Small mouth bass	1	0	0	0
Trout	1	0	0	0

TABLE 28 RELATIVE ABUNDANCE OF PHYTOPLANKTON

(value of 1 assigned to A)	A	B	C	D
Phytoplankton	1	0.9	1.3	2.8

Questions

1 What is the overall effect of the chemical plant on the life in the river?
2 Account for the high B.O.D. at station B.
3 What could cause the coliform count to be 3 at station B?
4 There is an increase in the abundance of phytoplankton at stations C and D, but a decrease at B. Why?
5 Account for the changes in the fish population.
6 Account for the reappearance at stations C and D of intolerant bottom fauna.
7 From the data provided, what would you expect the distribution and abundance of zooplankton to be at the four stations?
8 Account for the observed distribution of *Chironomus* and *Tubifex*.
9 Give a possible explanation for the presence of ammonia at A.
10 What is the nature of the major pollutants emitted by this plant?

8.4 A RIVER THAT PASSES THROUGH AN URBANIZED AREA

In 1965, a chemical and biological survey was made on the Upper Credit River in Ontario. Six stations were set up (Fig. 8-4). The stream at stations A, B, and C was 12–30 feet wide and 2–4 feet deep. The velocity of flow averaged 0.25 feet per second. Stations D, E, and F were of similar width, but only about half as deep. The velocity of flow rose to 1.5–2 feet per second. The occurrence of riffle areas was evident.

Station A was located above a sewage treatment plant at Orangeville. Station B was 1 mile downstream, just below the sewage treatment effluent site. Station C was 1.5 miles below B; station D was 1 mile below C. Station E was about 2 miles further downstream, and F was about 4 miles south of D.

TABLE 29 BOTTOM FAUNA

	A	B	C	D	E	F
Tubifex	0	280	14	24	1	2
Midge larvae	2	362	526	333	20	63
Leeches	1	3	0	4	5	0
Beetle larvae	0	21	1	44	0	0
Dragonfly nymphs	10	15	0	24	0	0
Caddisfly larvae	1	0	0	0	39	114
Mayfly nymphs	1	2	0	12	8	577
Stonefly nymphs	0	0	0	0	0	2

The results of the survey are summarized in Tables 29, 30, and 31.

TABLE 30 CHEMICAL ANALYSIS

	A	C	D	1 mile above E	1 mile below E
Dissolved oxygen (ppm)	6.5	1.8	4.7	8.4	8.7
B.O.D. (ppm)	1.3	2.8	4.6	3.7	3.1
T.S.S. (ppm)	16	10	21	21	46
T.D.S. (ppm)	335	430	368	349	350
Soluble phosphorus (ppm)	0.2	4.1	2.9	2.5	1.9
Total phosphorus (ppm)	0.2	6.2	3.6	3.3	2.6
Nitrogen, NH_3 (ppm)	0.1	2.1	0.2	0.3	0.1
Nitrogen, NO_2^- (ppm)	—	0.02	0.03	0.02	0.01
Nitrogen, NO_3^- (ppm)	0.19	0.15	0.69	0.23	0.19

Fig. 8-4
Upper Credit River.

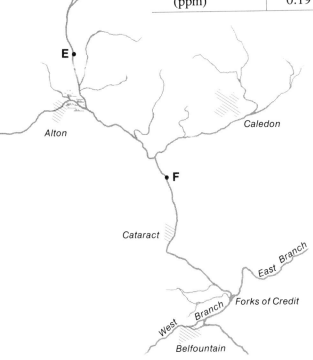

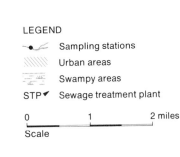

LEGEND
Sampling stations
Urban areas
Swampy areas
STP Sewage treatment plant

0   1   2 miles
Scale

TABLE 31 FISH CATCH

	A	B	C	D	E	F
Brook trout	0	0	0	0	0	5
White sucker	7	0	243	98	47	61
Red belly dace	10	0	82	2	7	0
Blacknose dace	0	0	0	2	39	36
Longnose dace	0	0	0	0	84	76
Chub	0	0	0	0	8	0
Blacknose shiner	88	0	7	0	0	0
Stickleback	8	0	4	36	2	0

Questions

1 What effect did the sewage treatment plant at Orangeville have on the water quality?
2 Station A is located above the effluent site and yet was low in pollution-sensitive organisms that usually inhabit a clean environment. What are possible explanations for this?
3 How would the bottom fauna at stations E and F compare to those at A if Orangeville were not on the Credit River?
4 At stations D, E, and F there is a significant decrease in the population of *Tubifex* and midge larvae, and an increase in the population of mayfly nymphs and caddisfly larvae. Why?
5 Explain the absence of all fish at station B.
6 Estimate the possible distribution of bacteria from station A to station F.
7 Compare carefully the distribution of *Tubifex* and midge larvae. Explain what you observe.
8 Why is the concentration of dragonfly nymphs so high at D?
9 Why would the B.O.D. increase between stations C and D? Normally a B.O.D. increase is accompanied by a D.O. decrease. Why did this not occur here?

8.5 THE "INFERNAL" COMBUSTION ENGINE

Engineering studies have reported the emissions recorded in Table 32 produced by the combustion of 1,000 gallons of gasoline.

Questions

1 How many miles is your family car driven during an average year? About how many gallons of gasoline are consumed during this period? Calculate the ap-

TABLE 32

Pollutant	Quantity (pounds)
Carbon monoxide	3,200
Organic vapors (mainly hydrocarbons)	200–400
Nitrogen oxides	20–75
Aldehydes	18
Sulfur compounds	17
Organic acids	2
Ammonia	2
Solid particles	0.3

proximate amount of each type of pollutant which your family car produces during an average year. What is the total quantity of pollutant emissions?

2 Why do car manufacturers use names such as Challenger, Wildcat, Tempest, Firebird, Demon, Mustang, and Javelin? What percentage of this type of automobile have engines requiring high-octane, leaded gasoline? Do they consume more gasoline per mile than the "standard" engines? How does pollutant emission compare? Are high compression engines more difficult to keep tuned? How does a poorly tuned engine contribute to air pollution?

3 (a) What are aldehydes? What effects do they have on humans?

(b) Under what circumstances might an automobile emit a higher than average percentage of particulate matter?

(c) Why are ranges given for the quantities of organic vapors and nitrogen oxides?

4 (a) Which pollutants are being reduced most by the emission control devices that are currently being installed on new cars?

(b) Do scientists predict any undesirable results from the use of these devices?

(c) Many cars are now being manufactured that run on non-leaded gasoline. This will significantly reduce the quantity of lead aerosols in the air. Will it have any undesirable effects?

5 There are cities in Asia with populations as high as 5 million in which few automobiles are found. Bicycles are the main form of transportation. In many European cities bicycles are also used by a large percentage of the population to go to and from work. Do you think bicycles could form the main method of transportation in your city (or the city closest to where you live)? Are automobiles an economic necessity in North America?

8.6 NATIONWIDE EMISSIONS OF AIR POLLUTANTS

Table 33 (page 246) summarizes the estimated nationwide (U.S.A.) emissions of air pollutants for 1968.

TABLE 33 ESTIMATED NATIONWIDE EMISSIONS, 1968
(Millions of tons per year)

Source	Carbon monoxide	Particulates	Sulfur oxides	Hydrocarbons	Nitrogen oxides
Transportation	63.8	1.2	0.8	16.6	8.1
Fuel combustion in stationary sources	1.9	8.9	24.4	0.7	10.0
Industrial processes	9.7	7.5	7.3	4.6	0.2
Solid waste disposal	7.8	1.1	0.1	1.6	0.6
Miscellaneous	16.9	9.6	0.6	8.5	1.7
Total	100.1	28.3	33.2	32.0	20.6

Questions

1 (a) Calculate the total amount of air pollutants produced in 1969.

(b) Of this total, what percentage was contributed by vehicles?

(c) For each of the 5 categories of pollutants, calculate the percentage that is contributed by vehicles.

(d) Opponents of expressway construction and of vehicular traffic in downtown areas quote the figure that you determined in (b) as evidence that the automobile is the main culprit in air pollution. Automobile manufacturers point out that the use of the total percentage figure is not realistic because some pollutants are more harmful than others. What do you think? Why?

2 (a) 24.4 million tons of sulfur oxides were produced by the burning of fuels in stationary sources. What percentage of the total is this?

(b) Coal combustion accounted for 20.1 million tons of sulfur oxides and fuel oil combustion for another 4.3 million tons. Stated differently, power plants produced 16.8 million tons, industrial plants 5.1 million tons, and space heating of homes and businesses, 2.5 million tons. Suggest steps that could be taken to lessen the amount of sulfur dioxide released into the air. Discuss the feasibility of your suggestions.

3 (a) What percentage of the total carbon monoxide emission was produced by transportation vehicles?

(b) Supporters of expressway construction in cities often state that automobiles produce more carbon monoxide when they are driven slowly. Therefore, they argue, carbon monoxide pollution can be reduced by building expressways to keep the traffic moving faster. Research appears to back up their opinions. In the United States vehicular traffic is almost evenly divided between rural and urban areas. Yet, 70% of the carbon monoxide produced by vehicles is found in urban areas where driving speeds are slower. Do you agree with those who support the construction of more expressways? Why?

4 (a) Nitrogen oxide emissions for 1970 totalled 23 million tons. What do you think were the main reasons for the increase over the 1968 value?

(b) There are over 940 power plants in the United States that burn fossil fuels. They produced 16.8 million tons of sulfur oxides and 4.0 million tons of nitrogen oxides. These large quantities of dangerous pollutants would be eliminated if the power plants were replaced by nuclear power plants. Discuss the feasibility of doing so.

5 Table 34 gives the populations of 5 cities and the number of tons of particulate matter falling on each of the cities per year.

TABLE 34

City	Population (metropolitan area)	Number of tons of particulate matter
Chicago, Ill.	7,225,000	600,000
Pittsburgh, Pa.	1,955,000	400,000
Philadelphia, Pa.	4,200,000	200,000
Los Angeles, Calif.	7,635,000	177,000
Denver, Colo.	1,035,000	33,000

(a) Calculate the amount of particulate matter that settles per year for each person in each of these cities.

(b) Rank these cities with the "dirtiest" city at the top.

(c) Examine carefully the sources of particulates given in Table 33. Account for the ranking in (b).

8.7 GEOGRAPHY, WEATHER, AND AIR POLLUTION

The state of North Carolina consists of three major geographical regions (Fig. 8-5):

Fig. 8-5
Geographical regions of North Carolina.

1)   The Coastal Plain. Low-lying, flat, and partly swampy outer tide-region. The inner region is gently rolling and well drained.

2)   The Piedmont Region. Well-rounded hills and low, rolling ridges. This area provides some moderately deep valleys, many of which are open-ended. Others form basins.

3)   The Mountain Region. A greatly dissected plateau bounded by two mountain chains. The North Carolina mountains are the most rugged land surface found in the eastern United States. The topography includes numerous valleys, many of which form closed basins.

The weather data recorded in Table 35 were collected at U.S. Weather Bureau Stations both in and bordering on the state of North Carolina.

Questions

1 Considering factors such as land surface features, prevailing winds, and moisture patterns, compare the three regions with respect to the following:
    (a) temperature variation from day to night;
    (b) the probability of temperature inversions;
    (c) the most likely season for temperature inversions;
    (d) the relative duration of occurring temperature inversions.
2 Why does the frequency of fog formation increase moving westward from the Coastal to the Mountain Region?
3 The general alignment of the main ridges of the Mountain Region is northeast-southwest. This corresponds to the prevailing wind flow of the Piedmont and Coastal Regions and the surface of the mountain ranges. Explain why the weather stations of the Mountain Region record wind directions which vary as much as 90° from this alignment.
4 Why is the average wind speed greater in the Coastal Plain Region?
5 To cope with the problem of air pollution, under the existing standards for emissions, the development sites of potential sources of pollutants should be carefully considered.
    (a) Compare the 3 regions with respect to suitability for the establishment of industrial communities.
    (b) Which areas in the Piedmont Region should be avoided?
    (c) Mining and timber-dependent industries are developing in the Mountain Region. Where would the related communities most likely settle? Why does this present a problem?
6 Suppose that each of the communities around the weather stations listed in Table 35 produced large quantities of airborne contaminants.
    (a) Which of these areas would be most severely affected? Why?

(b) Would the extent of this problem vary according to the season? Why?

(c) Outline the most probable distribution of these contaminants from each station.

TABLE 35 WEATHER CHARACTERISTICS OF NORTH CAROLINA

Region and Station	Coastal Plain Region			Piedmont Region			Mountain Region		
	Wil- mington N.C.	Hat- teras N.C.	Norfolk Va.	Char- lotte N.C.	Winston –Salem N.C.	Raleigh N.C.	Ash- ville N.C.	Bristol Tenn.	Green- ville S.C.
Prevailing wind direction									
Annual	SW	SSW	SW	SW	NE	SW	NW	W	NE
Winter	SW	NW	SW	SW	SW	SW	NW	W	SW
Spring	SW	SSW	SW	SW	SW	SW	NW	WSW	SW
Summer	SW	SSW	SW	SW	NE	SW	NW	NE	NE
Autumn	NE	NNE	NE	NNE	NE	NE	NW	E	NE
Approximate mean wind speed (mph)									
Annual	9.4	13.1	11.0	6.9	8.6	7.7	8.0	5.7	8.2
Winter	9.5	14.3	11.8	7.4	8.9	8.1	9.6	6.6	8.6
Spring	10.5	13.8	11.9	7.6	9.8	8.8	9.1	6.9	9.1
Summer	8.8	11.6	9.6	6.0	7.3	6.7	6.2	4.2	7.1
Autumn	8.5	12.7	10.7	6.5	8.5	7.1	7.4	5.3	7.8
Percent calms (winds 0–3 mph)									
Annual	16	1	13	28	13	25	42	42	14
Winter	14	1	10	28	14	21	35	37	11
Spring	12	1	10	21	7	19	54	34	10
Summer	18	1	18	32	16	28	50	47	18
Autumn	18	1	14	21	13	31	47	49	16
Annual number of days									
Heavy fog	11	9	16	20	34	18	56	40	16
Cloudy	103	112	121	126	120	118	121	160	129
Partly cloudy	129	111	115	116	107	116	129	117	110
Clear	133	142	129	123	118	131	115	88	126

8.8   DDT IN HUMAN BODY FAT

Chlorinated hydrocarbon insecticides have been widely used throughout the world. Aldrin, dieldrin, and DDT are three of these. They are now present in our food, water, and air. Since they degrade (break down) very slowly in nature, they are likely to be with us for many years to come.

DDT was the first of these insecticides to be marketed. Many countries still use it extensively in spite of frightening statistics regarding its toxic and accumulative properties. Human milk has been found to contain concentrations of DDT up to 5 ppm, yet it is illegal to sell cow's milk if the DDT concentration exceeds 0.05 ppm!

DDT tends to accumulate in the fatty tissues. Correlations have been discovered between DDT concentrations in these fatty tissues and the incidence of diseases such as hyper-

TABLE 36

Country	Year	Conc. of DDT (ppm)
Alaskan Eskimo	1960	3.0
Canada	1959–60	4.9
Canada	1966	3.8
Germany	1958–59	2.2
India (Delhi)	1964	26.0
Israel	1963–64	19.2
United Kingdom	1964	3.3
United States	1942	0
United States	1950	5.3
United States	1955	19.9
United States	1961–62	10.7
United States (national average)	1964	7.0
United States (New Orleans)	1964	10.3
United States (whites over 6 yrs of age)	1968	8.4
United States (non-whites over 6 yrs of age)	1968	16.7

tension, cerebral hemorrhage, softening of the brain, and some types of cancer.

Table 36 gives the average DDT concentration in human body fat for several countries and for several situations within one of these countries, the United States. The number of people tested in each case varied from 10 to 250.

Consult the references at the end of Section 5.5 for further information on DDT.

Questions

1   Account for the changes in the DDT concentration in human body fat of United States citizens between 1942 and 1962.

2   Why do residents of India and Israel have such high DDT concentrations?

3   Canadian agricultural techniques are, in general, quite similar to those of the United States. Why, then, is the average DDT concentration in the body fat of Canadians significantly less than that in citizens of the United States?

4   Account for the relatively low DDT concentration in the body fat of residents of the United Kingdom and Germany.

5   Eskimos generally eat a great deal of fat, yet the DDT concentration in their body fat is lower than the United States average. Why?

6   Account for the higher DDT concentration among the residents of New Orleans.

7   What causes the difference in the DDT concentration between whites and non-whites?

8   A very fat person who goes on a crash diet could expose himself to DDT poisoning. Why?

d) Further observations could be obtained by altering the rate at which smoke is introduced into the cylinder or by changing the nature of the smoke source. You could use punk, camphor, incense cones, and ammonium chloride (combined vapors from hydrochloric acid and ammonium hydroxide).

e) Disconnect the power source before dismantling any of the experimental apparatus.

Discussion

How effective is your model electrostatic precipitator? Does the efficiency of the model vary noticeably when the nature of the smoke or the rate of smoke production is altered? Are electrostatic precipitators used in any of the smoke stacks in your community? Design an experiment to determine *quantitatively* the efficiency of the model electrostatic precipitator.